Micellar Enhanced Ultrafiltration

Fundamentals & Applications

IIT Kharagpur Research Monograph Series

Micellar Enhanced Ultrafiltration

Fundamentals & Applications

SIRSHENDU DE
SOURAV MONDAL

CRC Press
Taylor & Francis Group
Boca Raton London New York

CRC Press is an imprint of the
Taylor & Francis Group, an **informa** business

CRC Press
Taylor & Francis Group
6000 Broken Sound Parkway NW, Suite 300
Boca Raton, FL 33487-2742

First issued in paperback 2019

ISBN-13: 978-1-4398-9568-9 (hbk)
ISBN-13: 978-0-367-38127-1 (pbk)

Library of Congress Cataloging-in-Publication Data

De, Sirshendu.
   Micellar enhanced ultrafiltration : fundamentals & applications / Sirshendu De, Sourav Mondal.
       p. cm. -- (Iit kharagpur research monograph series)
   Includes bibliographical references and index.
   ISBN 978-1-4398-9568-9 (hardback)
     1. Ultrafiltration. 2. Micelles. I. Mondal, Sourav. II. Title.

TP156.F5D27 2012
660'.284245--dc23   2011052566

Visit the Taylor & Francis Web site at
http://www.taylorandfrancis.com

and the CRC Press Web site at
http://www.crcpress.com

Contents

Preface

Pollution of the environment from the waste emerging from various industries is a burning social issue. In view of this, the norms for regulation of the level of toxicity in the industrial discharge are becoming more stringent nowadays. Continuous removal of organic and inorganic pollutants from aqueous streams with high efficiency and economy is therefore a big challenge to the scientific community. Traditional treatment processes are labor- and cost-intensive and require bigger space.

Membrane-based separation technology can offer an attractive alternative in this regard. Several membrane-based techniques, such as microfiltration, ultrafiltration, nanofiltration, and reverse osmosis, are currently used in a wide range of applications encompassing the textile, pulp and paper, sugar, chemical, pharmaceutical, biomedical, biotechnological, and food industries. Polluted water contains a large number of metal ions (Cu^{2+}, Cr^{3+}, Zn^{2+}, Sr^{2+}, Ca^{2+}, Pb^{2+}, Ni^{2+}, Mn^{2+}, Co^{2+}, As^{3+}, Fe^{2+}, etc.), anions (oxyanions, phosphates, ferrocyanide, etc.), and organic compounds like phenol, aniline, dyes, etc. A rate-governed separation process like reverse osmosis is very effective and efficient in removal of these contaminants. Since the operating pressure in reverse osmosis is quite high, this process is highly energy-intensive and requires large investment, rendering its limited practical application. Therefore, micellar-enhanced ultrafiltration, based on the principle of colloid and interfacial chemistry, is a better substitute. Micellar-enhanced ultrafiltration is a technology that employs surfactant micelles to solubilize inorganic and organic pollutants from the effluent stream and subsequently filters them to restrict the micelle-pollutant complex formed in the permeate stream. More than 90% removal efficiency along with high throughput can be attained by using a pollutant-specific surfactant (or a mixed surfactant system) and high-permeability membrane, depending on the charge and other physical properties of the contaminants. Since more open-sized membranes are used, this process involves less energy consumption. The amount of surfactant required is a minimal amount (only to obtain a critical micelle concentration), and the process is also economical. After the separation of the waste from the effluent/process stream, recovery of the surfactant is also possible. In fact, micellar-enhanced ultrafiltration is a viable technique that can remove almost all metal ions (heavy metals, lanthanides, radioactive, etc.) with a reasonably high efficiency and throughput by proper selection of surfactant and membrane.

Therefore, use of MEUF technology ensures lower operating pressure, less energy consumption, and removal of smaller-sized pollutants with higher throughput. This book presents state-of-the-art research on this topic with a detailed description of various aspects of this technology. Chapter 1 deals with the effects of pollution in water and its consequences. Comparison of various treatment processes and membrane technologies has been addressed. Fundamentals of ultrafiltration have been explained in Chapter 2. Different types of membrane modules and modeling approaches have been outlined in this chapter. Micellar-enhanced ultrafiltration involves principles of colloid chemistry. Theories of micelle formation, stability and dynamics of micelles, phenomena of counterion binding, and solubilization of organic pollutants are important aspects. These are covered in detail in Chapter 3. Selection of surfactants is of extreme importance in micellar-enhanced ultrafiltration, which is elucidated in Chapter 4.

Removal of inorganic (cations, anions, and their mixture) and organic pollutants by micellar-enhanced ultrafiltration has been described in depth in Chapters 5 and 6, respectively. Removal of metal ions encompassing group II to lanthanides has also been covered.

Various influencing factors regarding an increase in throughput and the operating problems associated herewith are discussed in Chapter 7. Considering the economy of the overall process, recovery and reuse of surfactants are essential. Technologies involving precipitation and other methods are elaborated in Chapter 8. Finally, a glimpse of other potential applications of this technology is illustrated in Chapter 9.

Since a complete book on such a topic does not exist today, the importance of it from an academic as well as industrial point of view is remarkably high. This book can be used as one of the texts for the course involving membrane technology and environmental science taught at the postgraduate level. Of course, this book can be an extremely useful reference for students and professionals in chemical engineering, environmental engineering, civil engineering, bioengineering, agricultural engineering, and industrial engineering.

We believe that this book would initiate some research interests and industrial development on the lines of green and clean technology. We hope that the readers will benefit from the applicability and significance of this technology through this book. Although we have put in our best efforts to organize all possible information regarding micellar-enhanced ultrafiltration, readers' comments and suggestions for improvement will be gratefully acknowledged.

Sirshendu De

Sourav Mondal

Series Preface

IIT Kharagpur had been a forerunner in research publications and this monograph series is a natural culmination. Empowered with vast experience of more than 60 years, the faculty now gets together with their glorious alumni to present bibles of information under the *IIT Kharagpur Research Monograph Series*.

Initiated during the Diamond Jubilee Year of the Institute, the Series aims at collating research and developments in various branches of science and engineering in a coherent manner. The Series, which will be an ongoing endeavour, is expected to be a source reference to fundamental research as well as to provide directions to young researchers. The presentations are in a format that these can serve as stand alone texts or reference books.

> The specific objective of this research monograph series is to encourage the eminent faculty and coveted alumni to spread and share knowledge and information to the global community for the betterment of mankind

The Institute

Indian Institute of Technology Kharagpur is one of the pioneering technological institutes in India and it is the first of its kind to be established immediately after the independence of India. It was founded in 18 August, 1951, at Hijli, Kharagpur, West Bengal, India. The IIT Kharagpur has the largest campus of all IITs, with an area of 2,100 acres. At present, it has 34 departments, centers, and schools and about 10,000 undergraduate, postgraduate, and research students with faculty strength of nearly 600; the number of faculty is expected to double within approximately five years. The faculty and the alumni of IIT Kharagpur are having wide global exposures with the advances of science and engineering. The experience and the contributions of the faculty, students and the alumni are expected to get exposed through this monograph series.

More on IIT Kharagpur is available at www.iitkgp.ac.in

Acknowledgments

It is a pleasure to thank all those who made this book possible and turned it into reality. It is a small endeavor to acknowledge the good wishes, blessings, and whole-hearted support of our near and dear ones, to those we are really indebted to.

About the Authors

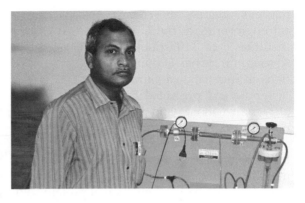

Dr. Sirshendu De is a professor in the Department of Chemical Engineering, Indian Institute of Technology, Kharagpur, India. He obtained his BTech (1990), MTech (1992), and PhD (1997) degrees from the Indian Institute of Technology, Kanpur. His major field of interest is membrane-based separation processes design, modeling, and fabrication of flat sheet and hollow fiber membranes. He has authored five books, nine book chapters, five patents, and more than 170 publications in national and international journals of repute. He has also transferred two technologies for commercialization. Other fields of research of Dr. De are adsorption, transport phenomena, modeling of flow through microchannels, etc. He has been the recipient of several prestigious awards, including the Shanti Swarup Bhatnagar Award in 2011 for fundamental contribution and innovation in basic science and technology (engineering science category). He also received the Herdillia Award in 2010 for excellence in basic research in chemical engineering, the VNMM Award in 2009 for excellence in innovative applied research, the Young Engineer Award from the Indian National Academy of Engineering in 2001 for excellence in engineering research, and the Amar Dye Chem Award in 2000 for excellence in chemical engineering research. He has been also a fellow of the Indian National Academy of Engineering, New Delhi, for his contribution in engineering and research.

Sourav Mondal received his undergraduate degree from Jadavpur University in chemical engineering in 2010. Presently he is pursuing his master's degree in chemical engineering at the Indian Institute of Technology, Kharagpur. He has 10 publications in international journals of repute and presented four papers in national and international conferences. He is also involved in projects for the development of a ceramic membrane module and molec-  ular dynamics-based simulation of protein solution.

1

Pollution and Importance of Micellar-Enhanced Ultrafiltration

The growth of human civilization has accelerated through the pace of the industrial revolution. The rise of human activity and industrial growth has led to the menace of environmental pollution to society. Pollution of water due to several reasons is one of the major concerns of the present-day world. The need for pure water is not only for drinking, but also for all sorts of activities. The contaminants to water from the major process industries, agriculture, households, etc., have social and ecological impact. In most of the cases, the need for proper effluent treatment before discharging into the environment is hardly realized. The rising limits on the regulation of toxicity demand more reliable, economical, and efficient methods of pollutant removal from the waste. The level of toxicity present in water is estimated by measuring the pH, chemical oxygen demand (COD), biological oxygen demand (BOD), total dissolved solids (TDS), total suspended solids (TSS), etc., of the polluted water.

Figure 1.1 shows that a large proportion of the cost (in fact, more than half of the total cost) is involved in prevention, treatment, and development of clean water resources to reduce the hazards of water pollution in India.[1] The average cost of cleanup (in terms of GDP) for China has been estimated at 2.6% of the country's GDP, for Mexico at 3.3%, up to 5% for countries in Eastern Europe, and less than 1–2% for other industrialized nations.[2] From an economic point of view, there is a strong necessity for the development of sustainable technology for water purification and treatment that should be highly efficient and cost-effective. Conventional separation processes such as chemical treatment, extraction, etc., are not very efficient in terms of the continuous mode of operation.

1.1 Sources of Water Pollution

The sources of wastewater can be primarily classified as agricultural products, industrial effluents, household wastes, natural calamities, etc. It can be emphasized that the amount of water consumed in these sectors is proportional to the amount of wastewater produced. The global water consumption in these sectors is presented in Figure 1.2.

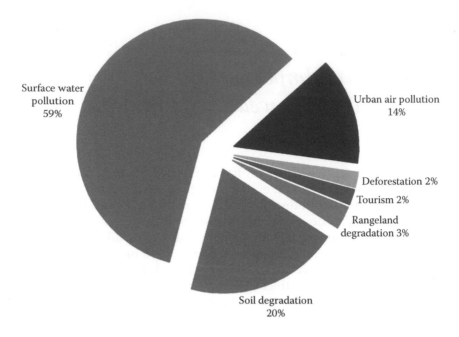

FIGURE 1.1 (See color insert.)
Proportion of cost (total cost U.S.$9.7 billion) due to pollution in India.[2]

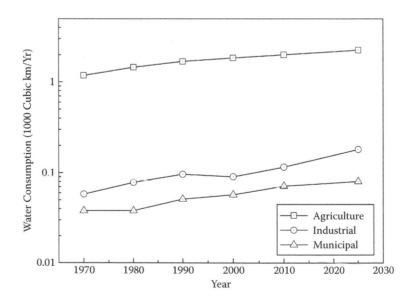

FIGURE 1.2
Global water consumption.

It is quite relevant to emphasize that global water pollution is directly related to the rate of global water consumption. The pertinent problem arises due to the fact that the water consumption rate has an ever-increasing trend in all three sectors (agriculture, industrial, and municipal), and hence there is a scarcity of the freshwater resources. Under such a global scenario, treatment of the wastewater becomes more than necessary. Figure 1.2 shows a projected estimate of the World Water Council in the year 2025.[3] It is noticeable that the industrial water consumption rate increases rapidly from the year 2000 compared to agricultural and municipal water consumption. Thus, the amount of polluted water this sector is emitting into the environment is significant. The global water consumption projection predicts that the amount of polluted water released from industrial streams is going to be double in the year 2025 from what it is today. Therefore, considering the ecological impact on the environment and society, we must now think of reliable and efficient pollutant (toxic/nontoxic) removal procedures and schemes.

1.1.1 Agricultural Wastewater

The pollution due to agricultural activities can be categorized as nonpoint pollution and point pollution. The point source includes wastes from animal and dairy products. The animal products typically consist of strong organic contents, high solids, nitrate, phosphate concentration, etc. Often significant amounts of bacteria, spores, and parasites are also found. The dairy waste mainly contains milk proteins, which are very conducive for growth of microorganisms since they are rich in nutritional value.

The nonpoint sources include sediment runoff from soil-washed fields. Excess sediment causes high levels of turbidity in water bodies, which can inhibit growth of aquatic plants, clog fish gills, and smother animal larvae.[4] Application of chemical fertilizers for enhancement of the crop yield leads to nutrient runoff that greatly affects the aquatic life[5] and increases the nonuniformity in soil fertility. Phenomena such as eutrophication affect the life of the phytoplankton. Pesticides are widely used by farmers to control plant pests and enhance production, but chemical pesticides can also cause water quality problems. Pesticides may appear in surface water due to direct application (e.g., aerial spraying or broadcasting over water bodies), runoff during rain storms, and aerial drift (from adjacent fields).

1.1.2 Industrial Effluent

The trend of biological oxygen demand (BOD) in India (1994–2004)[6] has been quantitatively presented in Figure 1.3. The figure shows the nature of BOD of the industrial effluents in India, specifically the BOD level in different wastewater sampling stations. The sampling stations reflect the collection of the samples in broad diversified areas (or different sources). A very positive feature is to be noted here: the number of samples labeled with desirable

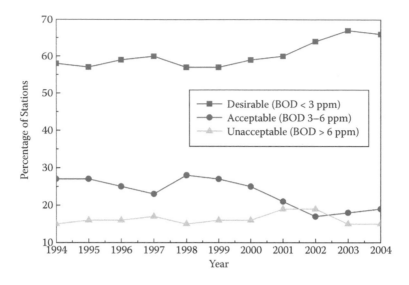

FIGURE 1.3 (See color insert.)
BOD of industrial effluent in India.

BOD limits are increasing as time progresses. This is a manifestation of the decrease in the number of sampling stations measuring acceptable BOD limits. However, there is hardly any change in the percentage of the number of stations with unacceptable BOD characteristics. This signifies that water resources of the areas (or sources) having unacceptable BOD limits do not improve in quality over time. This indicates that there is a strong need for a more viable and economic water treatment process.

1.1.3 Household Waste

The wastes from the house are typically kitchen wastes, washing outlets, etc., which contribute significantly to the total water pollution. Household waste contains a reasonable amount of detergents, surfactants, fats and oils, etc. (as reflected in COD), contributing to the higher organic content in these waste discharges.

Table 1.1 shows the typical concentration of the various contaminants generally present from household waste discharge. This finding has been reported by the United Nations Department of Technical Cooperation.

Various efforts to minimize waste production and alternate means for waste reduction, reuse, recycling, and disposal have been long studied. The proportion of different household waste out of total waste is represented in Figure 1.4.

TABLE 1.1

Major Constituents of Typical Domestic Wastewater[7]

Constituent	Concentration, mg/l		
	Strong	**Medium**	**Weak**
Total solids	1200	700	350
Dissolved solids (TDS)	850	500	250
Suspended solids	350	200	100
Nitrogen	85	40	20
Phosphorus	20	10	6
Chloride	100	50	30
Alkalinity	200	100	50
Grease	150	100	50
COD	250	500	1,000
BOD_5	300	200	100

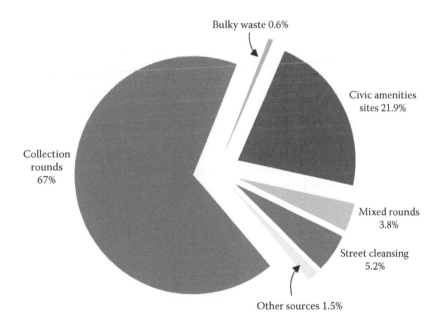

FIGURE 1.4 (See color insert.)
Estimated total household wastes for England and Wales in 1993–1994 (22.67 × 10⁶ kg/yr) by main outlets.[8]

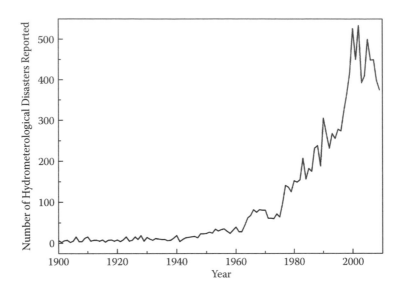

FIGURE 1.5
Global hydrological diasters.[9]

1.1.4 Natural Calamity

It is now a well-known fact that the natural disasters occurring on our planet have increased tremendously in the last couple of decades. The major impacts associated with these hydrological disasters, other than human casualties, include the contamination of groundwater with sewage water, leading to large-scale scarcity of safe drinking water in many parts of the world. Figure 1.5 clearly points out that the number of disasters occurring in the 1980s tripled by the year 2000.

The situation at present demands a simple but efficient means of water purification to provide pure water access to all of humankind.

1.2 Types of Pollutants and Sources

Typically pollutants can be classified as organic and inorganic chemicals present in polluted water. Inorganic pollutants constitute heavy metals, radioactive metals, nonmetals, etc. Organics are mostly comprised of organic manure and detergents. The sources of various pollutants are presented in Table 1.2.

TABLE 1.2

Different Sources of Organic and Inorganic Pollutants

Sources of Organic Contaminants	Sources of Inorganic Contaminants
Detergents	Acidity from industrial discharges (especially sulfur dioxide from power plants)
Disinfection by-products found in chemically disinfected drinking water, such as chloroform	Ammonia from food processing waste
Food processing waste, which can include oxygen-demanding substances, fats, and grease	Chemical waste as industrial by-products (textile effluent, leather processing units)
Insecticides and herbicides, a huge range of organohalides, and other chemical compounds	Fertilizers containing nutrients—nitrates and phosphates—which are found in storm water runoff from agriculture, as well as commercial and residential use
Petroleum hydrocarbons, including fuels (gasoline, diesel fuel, jet fuels, and fuel oil) and lubricants (motor oil), and fuel combustion by-products, from storm water runoff	Heavy metals from motor vehicles (via urban storm water runoff) and acid mine drainage
Volatile organic compounds (VOCs), such as industrial solvents, from improper storage; chlorinated solvents, which are dense nonaqueous phase liquids (DNAPLs), may fall to the bottom of reservoirs, since they don't mix well with water and are denser	Silt (sediment) in runoff from construction sites, logging, slash and burn practices, or land clearing sites
Various chemical compounds found in personal hygiene and cosmetic products	

1.3 Conventional Treatment Processes

1.3.1 Chemical (Inorganic) Methods of Treatment

The chemical treatment is relevant particularly in situations where the pollutant emitted has a potential interaction on a chemical reagent. The tendency of the pollutant to react and form an insoluble precipitate/colloidal suspension/nonobjectionable or eco-friendly product is desirable in this process. In general, six chemical processes can be used to remove substances from wastewater:

1. Reaction to produce an insoluble solid
2. Reaction to produce an insoluble gas
3. Reduction of surface charge to produce coagulation of a colloidal suspension

TABLE 1.3

Commonly Used Reagents for Chemical Treatment

Chemical Reagent	Application
Lime	Heavy metals, fluorides, phosphorous
Soda ash	Heavy metals
Sodium sulfide	Heavy metals
Hydrogen sulfide	Heavy metals
Phosphoric acid	Heavy metals
Fertilizer grade phosphate	Heavy metals
Ferric sulfate	Arsenic, sulfide
Ferric chloride	Arsenic, sulfide
Alum	Arsenic, fluoride
Sodium sulfate	Barium
Carbamates	Heavy metals

4. Reaction to produce a biologically degradable substance from a non-biodegradable substance

5. Reaction to destroy or otherwise deactivate a chelating agent

6. Use of chemical oxidation or reduction to produce nonobjectionable substances

The chemicals presented in Table 1.3 are commonly used for treating the targeted waste.[10]

The efficiency of pollutant removal by these chemical treatments is not very significant. Also, in cases of continuous removal of toxic components from the effluents, these treatments are not effective; moreover, removal of the precipitate or the neutralized complex formed requires further separation steps.

1.3.2 Biological (Organic) Methods of Treatment

It is a process in which the complex organic molecules that are present in the wastewater act as a feed to the microorganisms. Typically bacteria, fungi, and yeast microbes are allowed to grow over the waste and convert the pollutant as part of their protoplasmic constituents. Oxygen is required in either the dissolved molecular form or in the form of anions such as sulfate and nitrate. The end result is a decrease in the quantity of organic pollutants, and an increase in the quantity of microorganisms, carbon dioxide, water, and other by-products of microbial metabolism.

Bacteria and fungi are the primary converters. In a mature treatment system, these act the base of the hierarchical food pyramid. The protozoa, rotifers, and in some cases algae feed successively on these primary converters, thus forming a well-maintained continuous biological treatment system to manipulate

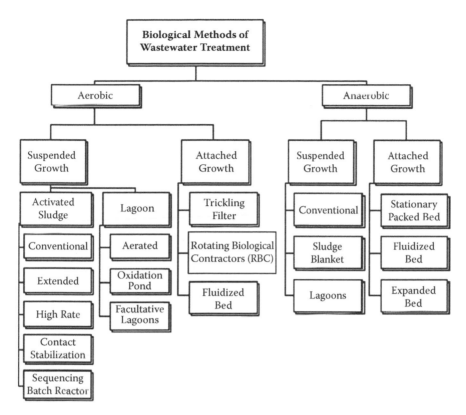

FIGURE 1.6
Different routes of biological waste water treatment.

the feeding rate, and the quantity of the microorganism is also controlled. Needless to say, the waste level is also regulated in an optimized fashion.[10]

The aerobic and anaerobic processes of biological treatment are presented in Figure 1.6. Biological processes are broadly categorized as aerobic and anaerobic. Both the aerobic and anaerobic forms of growth can be either in suspension or attached to the cultivated medium. Continuous processes for such growth patterns include trickling bed filter, rotating biological contactor (RBC), stationary packed bed in the case of anaerobic growth, fluidized bed, etc. The suspended growth involves use of lagoons, activated sludge, etc. (see Figure 1.6).

1.3.3 Common Effluent Treatment Plants (ETPs)

For the purpose of continuous wastewater treatment, the effluent treatment plants are set up in an all-effluent generation unit. The main unit in the ETP is the sedimentation tank or the slurry settler or thickener. The wastewater

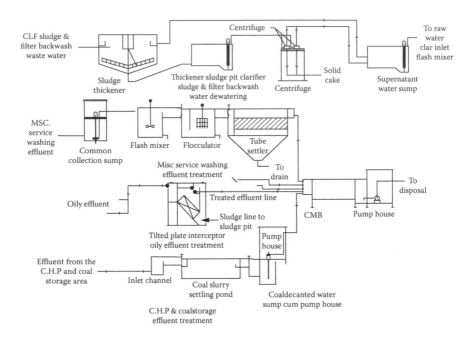

FIGURE 1.7
Schematic of a typical effluent treatment plant.

is fed into the settler and then subjected to posttreatment before being discharged into the environment to comply with the toxic limits.

Figure 1.7 shows an outline of the various processes in an effluent treatment plant. Different feed streams from various sources need to be pretreated accordingly. It may be emphasized that treated water from ETP does not meet the drinking water quality. It is for safer discharge as subsurface water.

1.3.4 Primary, Secondary, and Tertiary Treatment

Figure 1.8 shows the typical stages in primary, secondary, and tertiary processes of treatment of polluted water.[11]

1.3.4.1 Primary Treatment

The objective of primary treatment is the removal of settleable organic and inorganic solids by sedimentation, and the removal of materials that float (scum) by skimming. Approximately 25 to 50% of the present biochemical oxygen demand (BOD_5), 50 to 70% of the total suspended solids (SS), and 65% of the oil and grease are removed during primary treatment. Some organic nitrogen, organic phosphorus, and heavy metals associated with solids are also removed during primary sedimentation, but colloidal and dissolved

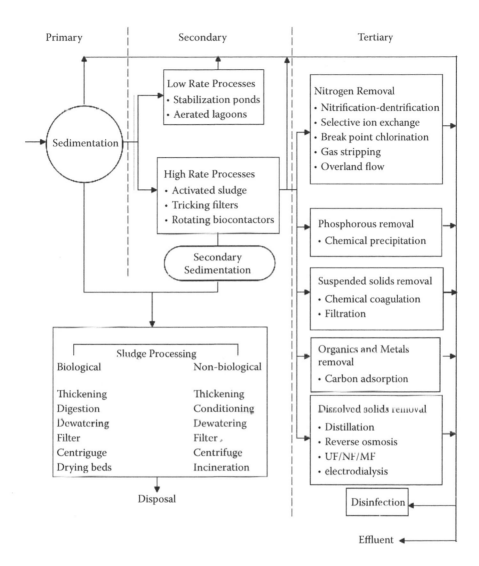

FIGURE 1.8
Different stages of conventional water treatment.

constituents remain unaffected. In many industrialized countries wastewater after primary treatment is considered sufficient for irrigating fields, but not for household use or human consumption.

1.3.4.2 Secondary Treatment

The objective of secondary treatment is the further treatment of the effluent from primary treatment to remove the residual organics and suspended solids.

In most cases, secondary treatment follows primary treatment and involves the removal of biodegradable dissolved and colloidal organic matter using aerobic biological treatment processes. Generally, aerobic biodegradation is performed on the primary treated wastewater by the action of microorganisms that metabolize the organic matter in the wastewater, thereby producing more microorganisms and inorganic end products (principally CO_2, NH_3, and H_2O). Several aerobic biological processes are used for secondary treatment differing primarily in the manner in which oxygen is supplied to the microorganisms and in the rate at which organisms metabolize the organic matter. Commonly used high efficient processes include activated sludge processes, trickling filters or biofilters, oxidation ditches, and rotating biological contactors (RBCs). These processes are capable of handling high concentrations of waste in a low reactor volume.

In the activated sludge process, the dispersed-growth reactor is an aeration tank or basin containing a suspension of the wastewater and microorganisms, the mixed liquor. The contents of the aeration tank are mixed vigorously by aeration devices that also supply oxygen to the biological suspension. Aeration devices commonly used include submerged diffusers that release compressed air and mechanical surface aerators that introduce air by agitating the liquid surface.

A trickling filter or biofilter consists of a basin or tower filled with support media such as stones, plastic shapes, or wooden slats. Wastewater is fed intermittently, or sometimes continuously, over the media. Microorganisms become attached to the media and form a biological layer or fixed film. Organic matter in the wastewater diffuses into the film, where it is metabolized. Oxygen is normally supplied to the film by the natural flow of air either up or down through the media, depending on the relative temperatures of the wastewater and ambient air.

Rotating biological contactors (RBCs) are fixed-film reactors similar to biofilters. Organisms are attached to support media. In the case of the RBC, the support media are slowly rotating disks that are partially submerged in flowing wastewater in the reactor. Oxygen is supplied to the attached biofilm from air when the film is out of the water; while submerged, oxygen is transferred to the wastewater by surface turbulence created by the disks' rotation.

1.3.4.3 Tertiary Water Treatment

Tertiary water treatment is the advanced stage of waste treatment that is employed when the desired water quality characteristics are not met by the above treatment procedures. Water for drinking and other household activities is often required to be processed through this stage. In the modern world perspective, apart from using an oxidation tank,[12] adsorption based on activated carbon,[13,14] ion exchange resins,[15,16] etc., provides means of tertiary water treatment. However, considering feasibility of process and

TABLE 1.4

Cost and Water Quality Comparison of Conventional and CMF Processes[20]

Water Quality Parameter	Influent	Conventional	CMF Treated
Turbidity, NTU	2–5	1	<0.1
Suspended solids, mg/l	5–10	2–3	<1
Total organic carbon, mg/l	10–12	8–10	8–10
Silt density index	>6	5–6	1–2
Bacteria, CFU/100 ml	10^5–10^6	3–4 log reduction	5–6 log reduction
Process performance (TOC, mg/l) (1999)		2 (blended 50% GAC + 50% RO)	0.3 (100% MF/RO)
Space required, $m^3/d.m^2$		21	106
Operating and Capital Costs			
Fixed Costs			
Capital costs ($/m³)	0.22	0.13	
O&M labor ($/m³)	0.04	0.02	
Replacement parts and supplies ($/m³)	0.01	0.02	
Subtotal fixed costs ($/m³)	0.27	0.17	
Variable Costs			
Chemical costs ($/m³)	0.09	0.03	
Sludge production and handling ($/m³)	0.06	0.003	
Power ($/m³)	0.02	0.02	
Subtotal variable costs ($/m³)	0.17	0.05	
Total fixed and variable costs ($/m3)	**0.43**	**0.22**	

application, membrane-based separation processes offer an attractive alternative (see Table 1.4). Membrane processes, which can offer a highly selective barrier to the water being processed, are far more robust to changes in feed water quality and can provide water of reliably high quality. The advantages of using a membrane separation process over conventional processes[17] are:

1. Separation is achieved without requiring a phase change, and is therefore more energetically efficient than distillation.
2. Little or no accumulation takes place in the process, which operates continuously under steady-state condition without necessitating regeneration cycles, unlike adsorptive separation processes.
3. No chemical addition is required, unlike conventional clarification involving the addition of chemical coagulants and flocculants.

TABLE 1.5

Major Applications of Various Membrane-Based Processes

Process	Usual Objective
Microfiltration (MF)	Removal of suspended solids, including microorganisms
Ultrafiltration (UF)	Removal of both large, dissolved solute molecules and suspended colloidal particles
Nanofiltration (NF)	(Selective) removal of multivalent ions and certain charged or polar molecules
Reverse osmosis (RO)	Removal of inorganic ions
Electrodialysis (ED) and dialysis	(Selective) extraction of ions from water or concentration
Gas transfer (GT)	Transfer of molecular gas into or out of water

A comparison of the conventional and continuous microfiltration (CMF) treatment processes based on the pilot work that has led to the operation of a 2,712 m³/d CMF/RO demonstration project[18,19] has been provided in Table 1.4.

Various filtration techniques, such microfiltration, nanofiltration, reverse osmosis, ultrafiltration, electrodialysis, etc., are useful in treatment procedures. Table 1.5 lists a few of the major application areas of these membrane-based techniques that are widely used in the present world.

1.4 Membrane-Based Separation Process

Membranes are defined as "an intervening phase separating two phases and/or acting as an active or passive barrier to the transport of matter between phases."[21]

The membranes and module sales in 1998 were estimated at more than U.S.$4.4 billion worldwide, shared by different applications.[22] The proportion of sales among different membrane processes in the worldwide market is represented in Figure 1.9. New applications in the emerging biotechnology field along with expanded use in wastewater treatment promise rapid

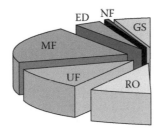
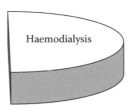

FIGURE 1.9
Qualitative representation of the global market share of the membrane industry.

growth in the coming years. Microfiltration, ultrafiltration, nanofiltration, and reverse osmosis are currently used in a very wide range of applications, including the textile, pulp and paper, sugar, chemical, pharmaceutical, biomedical, biotechnological, and food industries.

The textile industry produces a large amount of wastewater that is highly colored with high loading of inorganic salt. Different membrane processes were experimented on at pilot scale to verify the possibility of reusing textile wastewater. The pilot plant used sand filtration and UF as pretreatments for a membrane process of NF or reverse osmosis.[23] Cross-flow nanofiltration using a thin-film composite polysulfone membrane was used to recover the electrolyte solution and reject the color.[24] Nanofiltration of textile plant effluent for color removal (92–94%) and reduction in COD (94%) was attempted by Chakraborty et al.[25] Different types of MF, UF, and NF membranes were evaluated for permeate flux and their suitability in separating color and reducing COD, conductivity, total dissolved salts (TDS), and turbidity of textile effluent.[26] The combination of physicochemical treatment (coagulation/ flocculation) with NF or RO of textile effluent showed that a primary physicochemical treatment was necessary to limit membrane fouling.[27] New nanofiltration membranes have been developed by UV photografting. The grafted membranes were evaluated for the removal of six different dyes according to their charge in the aim of reusing the water in the process house. It was observed that the newly developed membranes showed acceptable performance in terms of both flux and rejection.[28]

The first full-scale reverse osmosis (RO) installation in the pulp and paper industry was put into operation in 1972 at Green Bay Packaging, Inc. The ultrafiltration process had been employed on a laboratory scale to remove suspended solids from dilute paper machine white water.[29] The treatment of white water from kraft corrugated waste by ultrafiltration and reverse osmosis has recently been investigated. A study had been undertaken on the utilization of inorganic membranes for the removal of color from effluents of the first caustic extraction in the pulp and paper production.[30] Satisfactory results were obtained for metal and COD removal from pulp and paper industry wastewater by coupling complexation with water-soluble polymeric ligands and the ultrafiltration process. In general, the complexation-ultrafiltration process increased the final quality of wastewater when compared to simple ultrafiltration.[31] Membrane technologies, namely, reverse osmosis and ultrafiltration, were applied on sectional drains of an integrated pulp and paper mill to test the suitability of these technologies for water recovery and chemical reclamation.[32] A laboratory fabricated stirred and rotating disk batch of UF cells was used in an attempt to minimize flux decline and thereby obtained enhanced flux for the treatment of black liquor obtained from sulfite pulping industries. An asymmetric cellulose triacetate membrane of 5,000 molecular weight cutoff was used for experimentations.[33]

Sugar processing is one of the most energy-intensive processes in the food and chemical industries. Therefore, membrane separation processes (MSPs) seem to find several applications there. The potential uses of membrane operations in both the cane and beet sugar industry are reviewed and discussed. Cost analysis shows that an increase of the membrane lifetime is required for ultrafiltration to be competitive to the conventional process, while reverse osmosis can be an alternative to enlargement of existing evaporators.[34] Applications of reverse osmosis in the treatment of sugar beat press water, recovery of salt from waste brine at a sugar decolorization plant, or microfiltration and ultrafiltration of raw sugar juice, as an alternative to chemical purification processes, have been cited.[35] Ultrafiltration has been used for clarification as well as for decolorization of raw brown sugar obtained from the Indian sugar industry.[36] A two-stage reverse osmosis system is investigated for preconcentrating the sugar syrup. The energy consumption is compared for conventional evaporation vs. reverse osmosis combined with evaporation. The obtained results represent several benefits, such as significant energy savings and no thermal loss of sugar using the RO system.[37]

Use of various membrane-based separations for recovery of polyvinyl alcohol,[38] phenol,[39] and clindamycin[40] from chemical industries is also reported. Purification of chlorine gas[41] and structured phospholipids produced by lipase-catalyzed acidolysis[42] is achieved by applying membrane-based separation processes.

The *United States Pharmacopeia* defines specifications and methods for production of water for injection. Ultrapure water systems in the pharmaceutical industry use reverse osmosis, ion exchange, and MF.[43] Though MF membranes can remove the microorganisms and particles above 0.2 µ, they cannot effectively remove pyrogens. It has been repeatedly demonstrated that a 10 kDa UF membrane can remove all pyrogens. Blood microfiltration is used to separate whole blood into plasma (which contains all the blood proteins and small molecular weight solutes) and a more concentrated cellular fraction.[44]

Ultrafiltration provides an ideal process for recovering the electropaint, eliminating the significant paint loss associated with the drag-out. In fact, the electrodeposition process would probably be impractical without the use of ultrafiltration.[45] Electrodeposition remains one of the most successful large-scale applications of UF. Blue whiting peptide hydrolysates are concentrated by UF as well as NF.[46]

UF systems for fruit juice clarification were initially developed in the 1970s, and are now used quite extensively throughout the industry.[47] The most significant application is in the production of clear apple juice,[48–50] although membrane filtration systems have also been used for the production of grape,[51] pear,[52] blood orange,[53,54] pineapple,[55] and lemon juices,[56] liquid foods,[57] and a number of important juice by-products (e.g., pectin and citrus peel extracts). Ultrafiltration is also used for the concentration of skim milk in the production of a variety of cheeses and dairy products.[58]

Separation by UF is pretty common in recovery of protein from cheese whey (cellulose acetate [CA], polysulfone, or polyamide membranes are suitable).[59]

UF and MF are very well suited to the processing of biological modules in the membrane bioreactors due to high surface area as well as the ability to obtain some separation/concentration of products and by-products.[60] UF fractionation of negatively charged proteins at pH 7, a-LA (target protein) and b-LG (contaminant protein) from whey protein concentrate, is investigated in order to reach a high selectivity of a-LA in the permeate.[61] Membrane (UF/MF) filtration is particularly well suited for the initial clarification of an antibiotic fermentation broth. One of the main advantages of membrane systems for the initial recovery of antibiotics from a fermentation broth is the ability to obtain very high yield using combined filtration and diafiltration processes.[62]

Drinking water must be essentially free from microorganisms, have sufficiently low concentrations of chemicals that are known to be injurious to health, and have an agreeable taste, odor, and color. Effective water treatment systems typically employ a wide range of different physical, chemical, and biological processes to produce water with the desired final purity. Ultrafiltration is now an accepted technology for filtration of drinking water. The world's biggest installation using UF is reported to be the Twin Oaks Valley Water Treatment Plant in San Diego, having a capacity of 100 million gallons per day. It is also a very useful technique of pretreatment of municipal wastewater before it is subjected to reverse osmosis.[59,62] Drinking water is prepared from seawater using a spiral wound and hollow fiber module with the help of reverse osmosis.[43]

1.5 Micellar-Enhanced Ultrafiltration

Polluted water contains a large number of metal ions (Cu^{2+}, Cr^{3+}, Zn^{2+}, Sr^{2+}, Ca^{2+}, Pb^{2+}, Ni^{2+}, Mn^{2+}, Co^{2+}, As^{3+}, Fe^{2+}, etc.), anions (oxyanions, phosphates, ferrocyanide, etc.), and organic compounds (like phenol, β-naphthol, p-nitrophenol, m-nitrophenol, catechol, o-chlorophenol, aniline, etc.). Most of the metal ions are toxic and potential environmental hazards. Organic compounds in these effluents undergo chemical as well as biological changes, consume dissolved oxygen, and destroy aquatic life. A rate-governed separation process like reverse osmosis (RO) is very effective and efficient in the removal of these contaminants. Since the operating pressure is very high and the permeability is very low, the process is highly energy-intensive and requires large investment. Therefore a membrane separation process, micellar-enhanced ultrafiltration (MEUF), is a better substitute considering the operating pressure requirement and usage of high-permeability membranes.[63] MEUF is a technology that employs surfactant micelles to solubilize inorganic and organic pollutants from the effluent stream.[64] MEUF is particularly effective for removal of single components such as Cd^{2+},[65] Mn^{2+},[66] Zn^{2+},[67] Cu^{2+},[68] Cr^{3+},[69] Pd^{2+},[70] Au^{3+},[71] etc., for simultaneous removal of Ni^{2+} and Co^{2+},[72] and Ni^{2+} and Zn^{2+}.[73]

Very small amounts of organic substances in water are effectively removed by MEUF.[74-79] Generally, dissolved organics are present along with heavy metals from industrial wastewater. Simultaneous removal of phenol or ortho-cresol and zinc or nickel ions using MEUF is reported by Dunn et al.[80] and Witek et al.[69]

Removal of anionic pollutants such as chromate, nitrate, permanganate, etc., using MEUF has been performed.[81-83] Separation of a mixture of anionic and cationic pollutants can be treated by a mixed micelle system. Micelles of cationic surfactants solubilize the anionic pollutants while the anionic surfactants solubilize the cationic pollutants.[84]

The presence of ionic surfactants in the treated effluent at higher concentration adversely affects the quality of treated water. On the other hand, the ionic surfactants are expensive. So, recovery and reuse of the surfactants are always preferred and desired in the MEUF system.[85,86]

References

1. Brandon, C., and Homan, K. 1995. The cost of inaction: Valuing the economic wide cost of environmental degradation in India. In *Asia Environment Division* (mimeo). World Bank, Washington, D.C.
2. Centre for Science and Environment. 1997. *Homicide by pesticides*. State of the Environment Series.
3. World Water Council. www. worldwater.council.org.
4. U.S. Environmental Protection Agency (EPA). 2005. *Protecting water quality from agricultural runoff*. Document EPA 841-F-05-001.
5. EPA. 2003. *National management measures to control nonpoint source pollution from agriculture*. Document EPA-841-B-03-004.
6. Rajaram, T., and Das, A. 2008. Water pollution by industrial effluents in India: Discharge scenarios and case for participatory ecosystem specific local regulation. *Futures* 40: 56–69.
7. UN Department of Technical Cooperation for Development. 1985.
8. Parfitt, J.P., Flowerdew, R., and Pocock, R. 1997. *A review of the United Kingdom household waste arisings and compositional data*. Report prepared under contract to the Department of the Environment, Wastes Technical Division EPG 7/10/21 CLO201, Environment Agency, London.
9. International disaster database. www.em-dat.net.
10. Woodard, F. 2000. *Industrial waste treatment handbook*. Oxford: Butterworth-Heinemann.
11. Levine, A.D., Tchobanoglous, G., and Asano, T. 1985. Characterization of the size distribution of contaminants in wastewater: Treatment and reuse implications. *J. Water Pollut. Control Fed.* 57: 805–816.
12. Weiner, R.F., and Matthews, R. 2003. *Environmental engineering*. Oxford: Butterworth-Heinemann (Elsevier Science).
13. Shih, T.C., Wangpaichitr, M., and Suffet, M. 2003. Evaluation of granular activated carbon technology for the removal of methyl tertiary butyl ether (MTBE) from drinking water. *Water Res.* 37: 375–385.

14. Snyder, S.A., Adham, S., Redding, A.M., et al. 2007. Role of membranes and activated carbon in the removal of endocrine disruptors and pharmaceuticals. *Desalination* 202: 156–181.
15. Dabrowski, A., Hubicki, Z., Podkoscielny P., et al. 2004. Selective removal of the heavy metal ions from waters and industrial wastewaters by ion-exchange method. *Chemosphere* 56: 91–106.
16. Rengaraj, S., Joo, C.K., Kim, Y., et al. 2003. Kinetics of removal of chromium from water and electronic process wastewater by ion exchange resins: 1200H, 1500H and IRN97H. *J. Hazard. Mater.* 102: 257–275.
17. Judd, S., and Jefferson, B. 2003. *Membranes for industrial wastewater recovery and re-use*. Oxford: Elsevier Ltd.
18. Dawes, T.M., Mills, W.R., McIntyre, D.F., et al. 1999. Meeting the demand for potable water in Orange County in the 21st century. The role of membrane processes. AWWA Membrane Technology Conference, Long Beach, CA.
19. OCWD. 1997. Status of research on the use of microfiltration for reclamation.
20. Durham, B., Bourbigot, M.M., and Pankratz, T. 2001. Membranes as pretreatment to desalination in wastewater reuse: Operating experience in the municipal and industrial sectors. *Desalination* 138: 83–90.
21. European Membrane Society. www.emsoc.eu
22. Nunes, S.P., and Peinemann, K.V. 2006. *Membrane technology in the chemical industry*. Weinheim: Wiley-VCH.
23. Marcucci, M., Nosenzo, G., Capannelli, G., Ciabatti, I., Corrieri, D., and Ciardelli, G. 2001. Treatment and reuse of textile effluents based on new ultrafiltration and other membrane technologies. *Desalination* 138: 75–82.
24. Tang, C., and Chen, V. 2002. Nanofiltration of textile wastewater for water reuse. *Desalination* 143: 11–20.
25. Chakraborty, S., Purkait, M.K., DasGupta, S., De, S., and Basu, J.K. 2003. Nanofiltration of textile plant effluent for color removal and reduction in COD. *Sep. Purif. Technol.* 31: 141–151.
26. Fersi, C., Gzara, L., and Dhahbi, M. 2005. Treatment of textile effluents by membrane technologies. *Desalination* 185: 399–409.
27. Suksaroj, C., Héran, M., Allègre, C., and Persin, F. 2005. Treatment of textile plant effluent by nanofiltration and/or reverse osmosis for water reuse. *Desalination* 178: 333–341.
28. Akbari, A., Desclaux, S., Rouch, J.C., Aptel, P., and Remigy, J.C. 2006. New UV-photografted nanofiltration membranes for the treatment of colored textile dye effluents, *J. Membr. Sci.* 286: 342–350.
29. Jönsson, A.S., and Wimmerstedt, R. 1985. The application of membrane technology in the pulp and paper industry. *Desalination* 53: 181–196.
30. Afonso, M.D., and de Pinho, M.N. 1991. Membrane separation processes in the pulp and paper industry. *Desalination* 85: 53–58.
31. Vieira, M., Tavares, C.R., Bergamasco, R., and Petrus, J.C.C. 2001. Application of ultrafiltration-complexation process for metal removal from pulp and paper industry wastewater. *J. Membr. Sci.* 194: 273–276.
32. Chakravorty, B., and Srivastava, A.S. 1987. Application of membrane technologies for recovery of water from pulp and paper mill effluents. *Desalination* 67: 363–369.
33. Bhattacharjee, C., and Bhattacharya, P.K. 2006. Ultrafiltration of black liquor using rotating disk membrane module. *Sep. Purif. Technol.* 49: 281–290.

34. Trägårdh, G., and Gekas, V. 1988. Membrane technology in the sugar industry. *Desalination* 69: 9–17.
35. Hinkova, A., Bubník, Z., Kadlec, P., and Pridal, J. 2002. Potentials of separation membranes in the sugar industry. *Sep. Purif. Technol.* 26: 101–110.
36. Hamachi, M., Gupta, B.B., and Ben Aim, R. 2003. Ultrafiltration: A means for decolorization of cane sugar solution. *Sep. Purif. Technol.* 30: 229–239.
37. Madaeni, S.S., and Zereshki, S. 2008. Reverse osmosis alternative: Energy implication for sugar industry. *Chem. Eng. Process.* 47: 1075–1080.
38. Porter, J.J. 1998. Recovery of polyvinyl alcohol and hot water from the textile wastewater using thermally stable membranes. *J. Membr. Sci.* 151: 45–53.
39. Han, S., Ferreira, F.C., and Livingston, A. 2001. Membrane aromatic recovery system (MARS)—A new membrane process for the recovery of phenols from wastewaters. *J. Membr. Sci.* 188: 219–233.
40. Zhu, A., Zhu, W., Wu, Z., and Jing, Y. 2003. Recovery of clindamycin from fermentation wastewater with nanofiltration membranes. *Water Res.* 37: 3718–3732.
41. Hägg, M.B. 2001. Purification of chlorine gas with membranes—An integrated process solution for magnesium production. *Sep. Purif. Technol.* 21: 261–278.
42. Vikbjerg, A.F., Jonsson, G., Mu, H., and Xu, X. 2006. Application of ultrafiltration membranes for purification of structured phospholipids produced by lipase-catalyzed acidolysis. *Sep. Purif. Technol.* 50: 184–191.
43. Porter, M.C. 2005. *Handbook of industrial membrane technology.* New Delhi: Crest Publishing House.
44. Zydney, A.L. 1995. Therapeutic apheresis and blood fractionation. In *Biomedical engineering handbook,* ed. A. Gronzino. Boca Raton, FL: CRC Publishing.
45. Wicks, Z.W. Jr. 1991. *Kirk-Othmer encyclopedia of chemical technology,* ed. M. Grayson and D. Eckroth, 352. New York: John Wiley & Sons.
46. Vandanjon, L., Johannsson, R., Derouiniot, M., Bourseau, P., and Jaouen, P.I. 2007. Concentration and purification of blue whiting peptide hydrolysates by membrane processes. *J. Food Eng.* 83: 581–589.
47. Blanck, R.G., and Eykamp, E. 1986. Fruit juice ultrafiltration. *AICHE Symp. Ser. Recent Adv. Separation Tech.* 82: 59–66.
48. Bélafi-Bakó, K., and Koroknai, B. 2006. Enhanced water flux in fruit juice concentration: Coupled operation of osmotic evaporation and membrane distillation. *J. Membr. Sci.* 269: 187–193.
49. Sheng, J., Johnson, R.A., and Lefebvre, M.S. 1991. Mass and heat transfer mechanisms in the osmotic distillation process. *Desalination* 80: 113–121.
50. He, Y., Ji, Z., and Li, S. 2007. Effective clarification of apple juice using membrane filtration without enzyme and pasteurization pretreatment. *Sep. Purif. Technol.* 57: 366–373.
51. Bailey, A.F.G., Barbe, A.M., Hogan, P.A., Johnson, R.A., and Sheng, J. 2000. The effect of ultrafiltration on the subsequent concentration of grape juice by osmotic distillation. *J. Membr. Sci.* 164: 195–204.
52. Warczok, J., Ferrando, M., López, F., and Güell, C. 2004. Concentration of apple and pear juices by nanofiltration at low pressures. *J. Food Eng.* 63: 63–70.
53. Galaverna, G., Silvestro, G.D., Cassano, A., et al. 2008. A new integrated membrane process for the production of concentrated blood orange juice: Effect on bioactive compounds and antioxidant activity. *Food Chem.* 106: 1021–1030.

54. Drioli, E., Jiao, B.L., and Calabro, V. 1994. Theoretical and experimental study on membrane distillation in the concentration of orange juice. *Ind. Eng. Chem. Res.* 33: 1803–1808.
55. de Barros, S.T.D., Andrade, C.M.G., Mendes, E.S., and Peres, L. 2003. Study of fouling mechanism in pineapple juice clarification by ultrafiltration. *J. Membr. Sci.* 215: 213–224.
56. Espamer, L., Pagliero, C., Ochoa, A., and Marchese, J. 2006. Clarification of lemon juice using membrane process. *Desalination* 200: 565–567.
57. Petrotos, K.B., and Lazarides, H.N. 2001. Osmotic concentration of liquid foods. *J. Food Eng.* 49: 201–206.
58. Kosikowski, F.V. 1996. Membranes separations in food processing. In *Membrane separation in biotechnology*, ed. W.C. McGregor. New York: Marcel Dekker. 201–254.
59. Dutta, B.K. 2007. *Principles of mass transfer and separation processes.* New Delhi: Prentice-Hall of India.
60. Cheryan, M., and Mehaia, M. 1986. Membrane bioreactors. In *membrane separation in biotechnology*, ed. W.C. McGregor, 255. New York: Marcel Dekker.
61. Lucas, D., Baudry, M.R., Millesime, L., Chaufer, B., and Daufin, G. 1998. Extraction of α-lactalbumin from whey protein concentrate with modified inorganic membranes. *J. Membr. Sci.* 148: 1–12.
62. Zeman, L.J., and Zydney, A.L. 1996. *Microfiltration and ultrafiltration principles and applications*, 544–551. New York: Marcel Dekker.
63. Purkait, M.K., DasGupta, S., and De, S. 2004. Resistance in series model for micellar enhanced ultrafiltration of eosin dye. *J. Colloid Interf. Sci.* 270: 496–506.
64. Myer, D. 1991. *Surface, interface and colloids—Principles and applications.* New York: VCH.
65. Ke, X., Guang-ming, Z., Jin-hui, H, et al. 2007. Removal of Cd^{2+} from synthetic wastewater using micellar-enhanced ultrafiltration with hollow fiber membrane. *Colloids Surf. A Physicochem. Eng. Aspects* 294: 140–146.
66. Juang, R.S., Xu, Y.Y., and Chen, C.L. 2003. Separation and removal of metal ions from dilute solutions using micellar-enhanced ultrafiltration. *J. Membr. Sci.* 218: 257–267.
67. Rahmanian, B., Pakizeh, M., and Maskooki, A. 2010. Micellar-enhanced ultrafiltration of zinc in synthetic wastewater using spiral-wound membrane. *J. Hazard. Mater.* 184: 261–267.
68. Liu, C.-K., and Li, C.-W. 2005. Combined electrolysis and micellar enhanced ultrafiltration (MEUF) process for metal removal. *Sep. Purif. Technol.* 43: 25–31.
69. Witek, A., Koltuniewicz, A., Kurczewski, B., Radziejowska, M., and Hatalski, M. 2006. Simultaneous removal of phenols and Cr^{3+} using micellar-enhanced ultrafiltration process. *Desalination* 191: 111–116.
70. Ghezzi, L., Robinson, B.H., Secco, F., Tiné, M.R., and Venturini, M. 2008. Removal and recovery of palladium(II) ions from water using micellar-enhanced ultrafiltration with a cationic surfactant. *Colloids Surf. A Physicochem. Eng. Aspects* 329: 12–17.
71. Akita, S., Yang, L., and Takeuchi, H. 1997. Micellar-enhanced ultrafiltration of gold(III) with nonionic surfactant. *J. Membr. Sci.* 133: 189–194.
72. Karate, V.D., and Marathe, K.V. 2008. Simultaneous removal of nickel and cobalt from aqueous stream by cross flow micellar enhanced ultrafiltration. *J. Hazard. Mater.* 157: 464–471.

73. Channarong, B., Lee, S.H., Bade, R., and Shipin, O.V. 2010. Simultaneous removal of nickel and zinc from aqueous solution by micellar-enhanced ultrafiltration and activated carbon fiber hybrid process. *Desalination* 262: 221–227.
74. Dunn Jr., R.O., Scamehorn, J.F., and Christian, S.D. 1985. Use of micellar-enhanced ultrafiltration to remove dissolved organics from aqueous stream. *Sep. Sci. Technol.* 20 257–284.
75. Dunn Jr., R.O., Scamehorn, J.F., and Christian, S.D. 1987. Concentration polarization effects in the use of micellar-enhanced ultrafiltration to remove dissolved organic pollutants. *Sep. Sci. Technol.* 22: 763–789.
76. Sabaté, J., Pujolà, M., and Llorens, J. 2002. Comparison of polysulfone and ceramic membranes for the separation of phenol in micellar-enhanced ultrafiltration. *J. Colloid Interface Sci.* 246: 157–163.
77. Syamal, M., De, S., and Bhattacharya, P.K. 1997. Phenol solubilization by cetyl pyridinium chloride micelles in micellar-enhanced ultrafiltration. *J. Membr. Sci.* 137: 99–107.
78. Adamczak, H., Materna, K., Urbanski, R., and Szymanowski, J. 1999. Ultrafiltration of micellar solutions containing phenols. *J. Colloid Interf. Sci.* 218: 359–368.
79. Gibbs, L.L., Scamehorn, J.F., and Christian, S.D. 1987. Removal of n-alcohols from aqueous streams using micellar-enhanced ultrafiltration. *J. Membr. Sci.* 30: 67–74.
80. Dunn Jr., R.O., Scamehorn, J.F., and Christian, S.D. 1989. Simultaneous removal of dissolved organics and divalent metal cations from water using micellar-enhanced ultrafiltration. *Colloids Surf.* 35: 49–56.
81. Baek, K., and Yang, J.W. 2004. Cross-flow micellar-enhanced ultrafiltration for removal of nitrate and chromate: Competitive binding. *J. Hazard. Mater.* B108: 119–123.
82. Purkait, M.K., DasGupta, S., and De, S. 2005. Simultaneous separation of two oxyanions from their mixture using micellar enhanced ultrafiltration. *Sep. Sci. Technol.* 40: 1439–1460.
83. Baek, K., Kim, B.K., and Yang, J.W. 2003. Application of micellar enhanced ultrafiltration for nutrients removal. *Desalination* 156: 137–144.
84. Das, C., Maity, P., DasGupta, S., and De, S. 2008. Separation of cation-anion mixture using micellar-enhanced ultrafiltration in a mixed micellar system. *Chem. Eng. J.* 144: 35–41.
85. Scamehorn, J.F., and Harwell, J.H. 1989. *Surfactant based separation processes.* Surfactant Science Series, Vol. 33. New York: Marcel Dekker.
86. Kim, H., Baek, K., Lee, J., Iqbal, J., and Yang, J.W. 2006. Comparison of separation methods of heavy metal from surfactant micellar solutions for the recovery of surfactant. *Desalination* 191: 186–192.

2

Fundamentals of Membrane Separation and Ultrafiltration

With increasing awareness of environmental pollution and stricter government regulation for the quality of industrial effluent, a zero discharge plant in any industrial production unit has become a thrust area. A greener, cheaper, and efficient treatment process in place of polluting and capital-intensive steps is desirable. Membrane-based processes can offer an attractive alternative in this regard.

2.1 Fundamentals

2.1.1 Nature of Separation Process

The separation processes, in general, are divided into equilibrium and rate-governed processes.[1] In equilibrium-governed separation processes, the product phases are in equilibrium with the inlet phases. Some of the popular equilibrium separation processes relevant industrially are distillation, absorption, adsorption, drying, etc. In such cases, the product streams are in equilibrium with feed streams. Thus, the process is quite slow and its performance is limited by the equilibrium of the streams at the operating temperature. In rate-governed separation processes, a difference of rate of physical transport of species leads the separation. The driving force of such transport is generally the gradient of chemical potential. The gradient of chemical potential is composed of four parts.[2] These are concentration gradient, pressure gradient, temperature gradient, and electrochemical potential gradient. The presence of these causes (one or more of such gradients) results in effects (the difference in separation and hence effects). Thus, if one can alter the driving forces of the system, the extent of separation and production rate can be controlled suitably. Most of the pressure-driven membrane processes fall in this category.

2.1.2 Advantages of Membrane-Based Process

The membrane-based systems have some distinct features over the conventional processes. These are listed below:[3–5]

- The separation can be achieved at room temperature. Thus, it is advantageous in case of processing of temperature-sensitive materials, like protein solution, fruit juice, etc., that can be denatured at higher temperature.
- There is no phase change during separation. Thus, capital- and energy-intensive equipment, like evaporators, etc., is not required.
- The separation is purely physical in nature and no chemicals are added to the system. This makes the system pollution-free and lowers the operating cost.
- The design of such a system is modular in nature and hence can easily be scaled up.
- The operation and maintenance are not labor-intensive.

2.1.3 Material of Construction

Membrane is a phase that separates two streams and allows selective transport of particular species through them. It can be polymeric or inorganic. Cellulose acetate, polyacrylonitrile, polysulfone, polyamide, etc., are some of the polymers to cast the membrane. On the other hand, alumina, zirconia, and ceramic materials are some of the inorganic materials. Inorganic membranes can withstand very high temperatures of the process streams. But, the controlling pore size is a major problem in inorganic membranes. Until now, achieving pore size up to the angstrom (10^{-10} m) level was not possible. On the other hand, in case of polymeric membranes, the pore size can reasonably be controlled in various sizes, ranging from microns (10^{-6} m) to angstrom levels.

2.1.4 Membrane Casting

Typically, a polymer is dissolved in an organic solvent of reasonable viscosity and cast on a nonwoven polyester fabric with thickness as low as 50 microns. A thin polymer film is produced when it is kept in a nonsolvent, like water, using the phase inversion technique. The pore size of the membrane is controlled by polymer-solvent composition, additives and their concentration, evaporation time, nature and duration in gelation bath, annealing temperature, etc. The details of membrane casting are available in the literature.[3]

2.1.5 Categorization of Processes

The pressure-driven, membrane-based separation processes are categorized in the following—the salient features of each presented below.

2.1.5.1 Reverse Osmosis (RO)

- Small solute particles to be separated
- Molecular weight < 100

- Pore size: 2–10 Å
- Pressure > 25 atm
- Permeation is main transport mechanism
- Example: Filtration of salt solution

2.1.5.2 Nanofiltration (NF)

- Particles to be separated with molecular weight: 200–1,000
- Pore size: 5–20 Å
- Pressure: 15–25 atm
- Particle retention of salts
- Example: Filtration of dyes, small molecular weight organics, etc.

2.1.5.3 Ultrafiltration (UF)

- Molecular weight of particles: 10^3– 10^5
- Pore size: 20–1,000 Å
- Pressure: 6–8 atm
- Transport mechanism: Convection (main) + diffusion
- Example: Filtration of protein, red blood cells, polymers, etc.

2.1.5.4 Microfiltration (MF)

- Molecular weight > 100 KDa
- Pore size > 1,000 Å
- Pressure: 2–4 atm
- Example: Filtration of clay solution, latex, paint, etc.

2.1.6 Transport Mechanism

2.1.6.1 Permeation

Solutes get dissolved in the membrane phase, diffused through it, and desorbed to the permeate side. The mechanism is prevalent in reverse osmosis and nanofiltration.

2.1.6.2 Knudsen Diffusion ($d/\lambda < 0.2$)

Single gaseous molecules diffuse under rarefied conditions so that the mean free path is longer than the pore diameter. The mechanism is observed in case of gas permeation through the membrane.

2.1.6.3 Convection (d/λ > 20)

Solutes and solvent get transported across the membrane as viscous flow. This mechanism occurs in ultrafiltration and microfiltration. In lower pore size ultrafiltration, both diffusion and convection come into play.

2.1.7 Characterization of Membranes

A membrane is characterized by the following parameters.

2.1.7.1 Observed Retention (Selectivity of Membrane)

This property indicates the extent of separation a membrane can produce with respect to the solute concentration in the feed. Thus, observed retention is defined as

$$R_o = 1 - \frac{C_p}{C_o} \qquad (2.1)$$

where C_p is solute concentration in permeate and C_o is solute concentration in feed.

If $R_o \rightarrow 1.0$, solute is completely retained by the membrane.

2.1.7.2 Real Retention

Real retention is a constant that defines the partition of the solute concentration across the membrane, i.e., between the membrane-solution interface and the permeate side. Since this definition is not masked by any physical phenomenon like deposition of solutes on the surface, etc., this indicates the true separation efficiency of the solute by the membrane.

$$R_r = 1 - \frac{C_p}{C_m} \qquad (2.2)$$

Here, C_m is the solute concentration in membrane interface. It may be mentioned here that since the membrane surface concentration of solute is always greater than the bulk concentration, real retention is always greater than the observed retention. For complete solute retention, $R_r = 1.0$.

2.1.7.3 Molecular Weight Cutoff (MWCO)

Molecular weight cutoff is a concept to characterize a membrane. In this case, generally retention of neutral solutes of various molecular weights is examined by conducting small-scale experiments. The typical operating

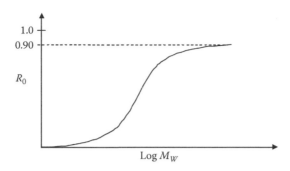

FIGURE 2.1
Typical molecular cutoff curve of a membrane.

conditions of these experiments are low transmembrane pressure drop, high turbulence, and low feed concentration. The observed retention values are then plotted against the molecular weight of the solutes in a semilog plot. The typical solutes are glucose (molecular weight 180), sucrose (molecular weight 342), various fractions of polyethylene glycol (molecular weights 200, 400, 600, 1,000, 1,500, 2,000, 4,000, 60,000, 10,000, 30,000), dextran (molecular weights 40,000 and 150,000), etc. The molecular weight at 90% solute retention indicates the molecular weight cutoff of the membrane. The molecular weight cutoff curves are shown in Figures 2.1 and 2.2.

Molecular weight cutoff curves may be sharp or diffused. If the retention curve raises sharply to a 90% level over a small span of molecular weight regime, then the cutoff curve is called a sharp cutoff curve. If the retention curve rises over a wide span of molecular weight region, it is a diffused cutoff curve. One has to have an accurate control over the operating conditions to achieve the sharp cutoff membranes. In fact, most of the commercial membranes are of the diffused type.

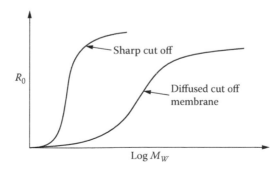

FIGURE 2.2
Sharp and diffused molecular cutoff curves.

2.1.7.4 Membrane Permeability (L_p)

This parameter shows how porous the membrane is. If L_p is more, then the membrane is more porous. Mathematically, L_p is defined as $L_p = \frac{J^0}{\Delta P}$, where J^0 is the pure water flux and ΔP is the transmembrane pressure drop. This concept is elaborated in detail in the subsequent sections.

2.1.8 Estimation of Retention and Permeability

2.1.8.1 Retention

Observed retention (R_o): Estimated by direct experimental measurement.

Real retention (R_r): One has to conduct batch experiments at high stirring speed, low feed concentration, and low operating pressure. In that case, it is assumed that there is no formation of concentrated solute layer over the membrane surface and in the absence of a polarized layer, observed retention is almost the same as real retention.[6]

2.1.8.2 Permeability

Membrane permeability is measured by distilled water runs. Experiments are conducted using distilled water at various transmembrane pressure drop values. At various pressure drops, the water flux is measured. A plot of permeate flux vs. operating pressure would be a straight line through the origin, as shown in Figure 2.3. The slope of this curve indicates the permeability (L_p) of the membrane. It may be noted here that the permeability of a membrane is its pressure history only. In other words, permeability of the membrane is independent of turbulence (stirring speed or cross-flow velocity) in the flow channel. The unit of permeability is $\frac{m}{Pa.s}$.

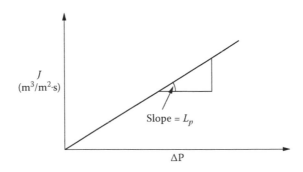

FIGURE 2.3
A typical flux vs. pressure plot for distilled water as feed.

2.2 Membrane Modules

The practical equipment where the actual membrane-based separation occurs is known as membrane modules. The basic aim of development of these modules is to provide a maximum membrane area in a relatively smaller volume, so that the permeate flux, i.e., the productivity of the system, is maximum. These membrane modules are of four types: (1) plate and frame modules, (2) hollow fiber modules, (3) spiral wound modules, and (4) tubular modules. Each of these modules is described below.

2.2.1 Plate and Frame Modules[3,7,8]

The heart of a plate-frame module is the support plate that is sandwiched between two flat sheet membranes. The membranes are sealed to the plate, either gaskets with locking devices, glue, or directly bonded. The plate is internally porous and provides a flow channel for the permeate that is collected from a tube on the side of the plate. Ribs or grooves on the face of the plate provide a feed side flow channel. The feed channel can be a clear path with channel heights from 0.3 to 0.75 mm. The higher channel heights are necessary for high-viscosity feeds; reduction in power consumption of 20 to 40% can be achieved by using a 0.6 mm channel compared to a 0.3 mm channel. Alternatively, retentate separator screens (20- or 50-mesh polypropelyne) can be used. Commercial plate-frame units are usually horizontal with the membrane plates mounted vertically. They can be run with each plate in parallel plates in two or three series. Modules are also available as preformed stacks of up to 10 plates. A typical plate and frame module is shown in Figure 2.4a.

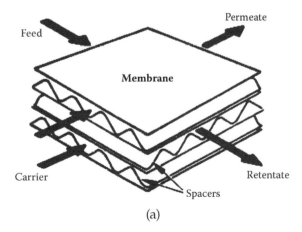

(a)

FIGURE 2.4
(a) A plate and frame module. (b) Picture of tubular membrane module. (c) Pictures of hollow fiber modules with end cap. (d) Flow pattern in hollow fiber module. (e) Flow pattern in spiral wound module.

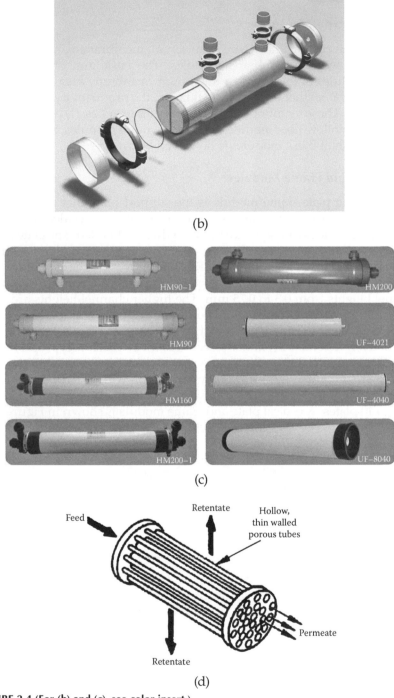

(b)

(c)

(d)

FIGURE 2.4 (For (b) and (c), see color insert.)
(Continued).

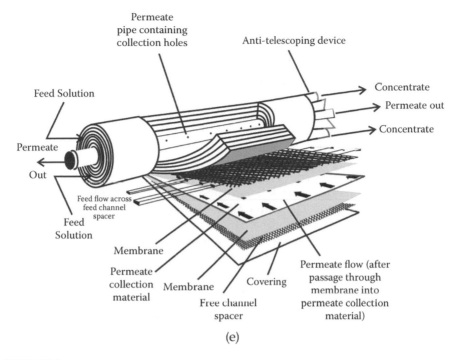

Permeate pipe containing collection holes

Anti-telescoping device

Feed Solution

Concentrate

Permeate out

Concentrate

Permeate

Out

Feed flow across feed channel spacer

Feed Solution

Membrane

Permeate collection material

Membrane

Covering

Free channel spacer

Permeate flow (after passage through membrane into permeate collection material)

(e)

FIGURE 2.4
(Continued).

2.2.2 Tubular Modules[3,7,8]

In such modules, the membrane is cast on the inside surface of a porous tube. Tubular membranes operate in tangential, or cross-flow, design where process fluid is pumped along the membrane surface in a sweeping type action. The feed solution is pumped through the center of the tube at velocities as high as 6 m/s. These cross-flow velocities minimize the formation of a concentration polarization layer on the membrane surface, promoting high and stable flux and easy cleaning, especially when the objective is to achieve high suspended solids in the MF, UF, or NF concentrate. Permeate is driven through the membrane to be directed out of the system or back into the process, depending on the application. There are many advantages in tubular membrane configurations. Besides their rugged construction, they have a distinct advantage of being able to process high suspended solids, and concentrate product successfully and repeatedly to relatively high end-point concentration levels without plugging. A common objective of an end-of-pipe waste treatment UF system is to reduce waste volume as much as possible to reduce concentrate hauling costs. For juice clarification applications, tubular membrane systems produce the greatest yields and the highest final suspended solids concentration levels. Tubular MF, UF, and NF systems do not require significant prefiltration. Some tubular products have the ability to be mechanically cleaned with sponge balls. Sponge balls can be used in process, and are also used to enhance chemical cleaning by reducing time and cleaning chemicals. Tubular membranes are ideally suited to treatment of metalworking oily waste, wastewater minimization and recovery from industrial processes, juice clarification, treatment of pulp and paper industry waste, etc. Tubular membranes typically have life up to 2 to 10 years. Figure 2.4b shows some tubular membranes.

2.2.3 Hollow Fiber Module[3,7,8]

In a hollow fiber module, lots of hollow fibers (each fiber is a tubular module) are kept in a large pipe. Geometry allows a high membrane surface area to be contained in a compact module. This means large volumes can be filtered, while utilizing minimal space, with low power consumption. Hollow fiber membranes can be designed for circulation, dead end, and single-pass operation. Some of the many hollow fiber membrane filtration applications include potable water treatment, juice clarification, wine filtration, dairy processing, etc. The advantages of such modules include reduction in space requirement, lowering in labor cost, lowering in chemical cost, delivery of high-quality product water, etc. Hollow fiber membranes offer the unique benefits of high membrane packing densities, sanitary designs, and due to their structural integrity and construction, can withstand permeate backpressure, thus allowing flexibility in system design and operation. Most hollow fiber products are available in (1) 1 inch diameter laboratory test cartridges, ranging up to 10 inches in diameter for commercial products; (2) standard commercial cartridge lengths of 25, 43, 48, 60, and 72 inches; (3) nominal separation ranges from 0.2 micron down to 1,000 MWCO, (4) fiber inside diameters from 0.02 inch (0.5mm) up to 0.106 inch (2.7 mm), and (5) various materials of construction, including polysulfone and polyacrylonitrile.

Figure 2.4c shows some hollow fiber cartridges of 5, 8, and 10 inches in diameter with endcaps.

Benefits of hollow fiber membranes include (1) controlled flow hydraulics, (2) tangential flow along the membrane surface limits membrane fouling, (3) membranes can be backflushed to remove solids from the membrane's inside surface, thus extending the time between two chemical cleaning cycles, (4) high membrane packing density resulting in high throughput, and (5) modular in structure so that future extension of the plant becomes easier. The flow pattern in a typical hollow fiber module takes place as shown in Figure 2.4d.

2.2.4 Spiral Wound Module[3,7,8]

In a spiral wound membrane, the membrane is cast as a film onto a flat sheet. Membranes are sandwiched together with feed spacers (typical thickness

0.03 to 0.1 inch) and permeate the carrier. They are sealed at each edge and wound up around a perforated tube. The module diameter ranges from 2.5 to 18 inches and the length varies from 30 to 60 inches. The typical cross section of the spiral wound module is shown in Figure 2.4e.

The application of a spiral wound module includes seawater desalination, brackish water treatment, potable water treatment, dairy processing, electrocoating paint recovery, protein separation, whey protein concentration, etc.

Therefore, it can be identified that the modeling of plate and frame and a spiral wound module can be done by considering the flow through a rectangular channel. On the other hand, tubular and hollow fiber modules are done by considering flow through a tube.

2.3 Operational Problems

2.3.1 Concentration Polarization

Accumulation of solute particles over the membrane surface establishes a development of a thin boundary layer of solutes, near the surface. This phenomenon is known as concentration polarization.[9–11] This phenomenon gives rise to the following:

1. Decline in driving force as the osmotic pressure at the membrane-solution interface increases with concentration
2. Formation of a gel type layer over the membrane surface
3. Adsorption of solutes within the membrane pore and matrix, therefore lowering the membrane permeability
4. Increase in concentration near the membrane surface, making the solution more viscous and offering more resistance against the solvent flux

The concentration polarization phenomenon is schematically shown in Figure 2.5.

Therefore, concentration polarization is very detrimental to the performance of a membrane system. It leads to a decrease in throughput of the process and deteriorates the quality of the permeate stream by adversely affecting the solute transport through the membrane. This phenomenon initiates the fouling of the membrane.

2.3.2 Membrane Fouling

Membrane fouling is of two types: reversible and irreversible.

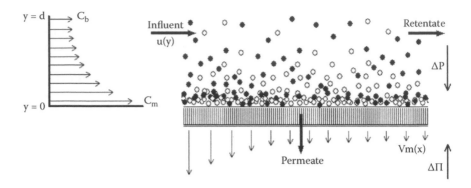

FIGURE 2.5
Concentration polarization in a membrane channel.

2.3.2.1 Reversible Fouling

This type of fouling is cleaned by adopting an appropriate cleaning protocol, like membrane washing. After cleaning, membrane permeability is restored.

2.3.2.2 Irreversible Fouling

In this case, membrane pores are blocked permanently and cannot be removed, even after appropriate cleaning. Some fraction of permeability of the membrane is lost permanently.

2.4 Modeling of Concentration Polarization

As discussed earlier, concentration polarization has to be quantified in order to analyze the system performance. This is a critical issue and has been an active field of research for several decades. As shown in Figure 2.5, there are two domains of flow; the first is in the channel outside the membrane surface, and the second is within the membrane pores. Modeling approaches can be classified based on the complications involved in the modeling in the following categories: (1) First-generation models are simplistic in nature and are mostly one-dimensional and hence deal with the average quantities. (2) Second-generation models are more detailed. The transport phenomena in the flow channel are considered to be two-dimensional. The transport law through the porous membrane is still averaged, but coupled with the boundary condition of the governing equation of the problem outside the membrane in the flow channel. (3) Third-generation models are the most complicated. The detailed two/three-dimensional transport phenomena in the flow channel are coupled with a detailed one/two-dimensional pore flow modeling, and the coupling appears at every axial location with the boundary condition of the problem associated in the flow channel. These three types of modeling approaches are discussed in brief.

2.4.1 First-Generation Models

As shown in Figure 2.5, the deposited solutes over the membrane surface create a thin concentration boundary layer. In this category of the model, it is assumed that the thickness of the concentration boundary layer is constant. The associated solute fluxes are (1) convective flux toward the membrane due to transmembrane pressure drop, (2) diffusive flux away from the membrane, and (3) convective flux away from the membrane toward the permeate side. At the steady state, near the membrane surface, the sum of all the fluxes toward the membrane is zero. This leads to the following solute balance equation:[9,12–14]

$$v_w c - v_w c_p + D \frac{dc}{dy} = 0 \tag{2.3}$$

Integrating the above equation across the mass transfer boundary layer thickness, the governing equation of the flux is obtained as

$$v_w = \left(\frac{D}{\delta}\right) \ln\left(\frac{c_m - c_p}{c_0 - c_p}\right) = k \ln\left(\frac{c_m - c_p}{c_0 - c_p}\right) \tag{2.4}$$

In the above equation, v_w is the permeate flux, k is the mass transfer coefficient, and c_m, c_p, and c_0 are solute concentrations at the membrane surface, in the permeate, and in the bulk, respectively. The mass transfer coefficient is estimated from the following equations depending on the channel geometry and flow regimes.

In a rectangular channel, the mass transfer coefficient is estimated as:[10–11] For laminar flow (Leveque's equation):

$$Sh = \frac{k d_e}{D} = 1.85 \left(\mathrm{Re}\, Sc \frac{d_e}{L} \right)^{\frac{1}{3}} \tag{2.5a}$$

For turbulent flow (Dittus-Boelter equation):

$$Sh = 0.023 (\mathrm{Re})^{0.8} (Sc)^{0.33} \tag{2.5b}$$

In the case of flow through the tube, the mass transfer coefficient is estimated for laminar flow (Leveque's equation):[10,11]

$$Sh = \frac{kd}{D} = 1.62 \left(\text{Re} \, Sc \, \frac{d_e}{L} \right)^{\frac{1}{3}} \tag{2.5c}$$

and for the turbulent flow, it is calculated from Equation (2.4). Now, the transport equation in the flow channel, Equation (2.2), must be coupled with the transport law through the porous membrane surface. It is expressed as Darcy's law:[3–5]

$$v_w = L_P (\Delta P - \Delta \pi) \tag{2.6}$$

where $\Delta \pi$ is the osmotic pressure difference across the membrane. The osmotic pressure, being a colligative property, is a strong function (in fact, ever increasing function) with concentration. Osmotic pressure is also inversely proportional to the molecular weight of solute. Therefore, it is quite significant for the solutes of lower molecular weights, like salts, dyes, etc. For charge-charge interaction, it also becomes important for the protein molecules. For salts and lower molecular weight solutes, it is a linear function of concentration, and for polymers, proteins, and higher molecular weight solutes, it is a nonlinear function of concentration. It can be noted that in Equation (2.2), there are three unknowns: v_w, c_m, and c_p. Osmotic pressure difference can be written in terms of concentration at the membrane surface as

$$\Delta \pi = \pi_m - \pi_p = a_1 (c_m - c_p) + a_2 \left(c_m^2 - c_p^2 \right) + a_3 \left(c_m^3 - c_p^3 \right) \tag{2.7}$$

The relation between c_m and c_p defines the partition coefficient across the membrane phase, between upstream and downstream sides of the membrane, known as real retention. This is a constant property, as defined in Equation (2.2).

Therefore, Equation (2.7) can be written in terms of single parameter c_m using Equation (2.2). Now the system variables are reduced to two, namely, c_m and v_w, instead of three. These two can be obtained by solving Equations (2.4) and (2.6) using an iterative algorithm like the Newton-Raphson technique. The above model is known as the classical film theory or osmotic pressure controlling model. This model is valid for the solutes of low molecular weights. Another variant of this model is the gel/cake controlling model. Typically, a gel/cake controlling phenomenon is observed for the solutes having high molecular weights. A gel on the membrane surface can be formed in two ways. First, the filtration can be osmotic pressure controlling initially. This leads to an enhancement of solute concentration on the membrane surface axially downstream of the flow channel. At one stage, the surface concentration of the solute exceeds the solubility limit of the solute at that temperature. Therefore, solutes can deposit over the membrane surface after phase separation. Second, a gel or a viscous layer forms over the membrane surface from the very beginning of the filtration process. This happens for some special solutes, which are well known for gel formation, like pectin, polyvinyl alcohol, high molecular weight polymers, proteins, etc. The osmotic pressure of these is negligible. Therefore, the gel layer is assumed to be a thick, highly viscous porous layer that forms over the membrane surface. For the gel controlling filtration case, it is assumed that the permeate stream does not contain the solutes, as the gel-forming solutes are much bigger in size with respect to the membrane pores and completely retained by the membrane simply by size exclusion. The system performance in terms of throughput is estimated from Equation (2.4) by putting c_p equal to zero and replacing c_m by c_g, the gel layer concentration. Gel layer concentration for a particular solute is constant, irrespective of selection of membrane. It is measured from a separate set of controlled and accurate experiments.[16]

2.4.1.1 Shortcoming of the Above Method

This method is associated with the following shortcomings:

1. The mass transfer boundary layer is assumed to be fully developed, whereas the entrance length required for the mass transfer boundary layer to be fully developed is substantial.
2. Variation of physical properties with concentration, like diffusivity, viscosity, etc., is not considered.
3. Mass transfer coefficients are used as obtained from heat-mass transfer analogies applicable for impervious conduits.

However, the film theory or osmotic pressure model presents a simplistic and quicker approach to quantify the system performance.

2.4.2 Second-Generation Models

In order to circumvent the limitations mentioned above, a two-dimensional model is formulated so that the developing mass transfer boundary layer can be accommodated.

The governing equation of the solute mass balance in the differential form within the mass transfer boundary layer is written as[16]

$$u\frac{\partial c}{\partial x}+v\frac{\partial c}{\partial x}=D\frac{\partial^2 c}{\partial x^2} \tag{2.8}$$

The velocity profile in the *x*-direction is assumed to be fully developed, as it requires a very short length (in the order of a few centimeters) and is given as

$$u = \frac{3}{2}u_0\left[1-\left(\frac{y-h}{h}\right)^2\right]$$

(2.9)

Within the thin mass transfer boundary layer, y^2/h^2 can be neglected and the above equation can be approximated as[17]

$$u = 3u_0\frac{y}{h}$$

(2.10)

Since the thickness of the boundary layer is extremely small, it is assumed that the *y*-component velocity remains almost invariant across it.[17,18]

$$v = -v_w(x)$$

(2.11)

The boundary conditions for Equation (2.8) are

$$\text{At } x = 0, c = c_0$$

(2.12)

$$\text{At } y = \infty, c = c_0$$

(2.13)

$$\text{At } y = 0, D\frac{\partial c}{\partial y} + v_w(c - c_p) = 0$$

(2.14)

The governing equation (2.8) along with its boundary conditions, Equations (2.12) to (2.14) in conjunction with Equation (2.6), can be solved using either (1) a similarity solution method, (2) an integral method, or (3) a numerically finite difference scheme. Details of these methods are available in the literature.[16–25] Finally, after solution, the profiles of state variables, v_w and c_m, are obtained as a function of *x*, and the length averaged (using Simpson's method) permeate flux and membrane surface and permeate concentration can be computed. Still this model suffers some shortcomings. The major one is the characterization of solute transport through the membrane pores. It is quantified by the following methods: (1) Define a constant coefficient (real retention) across the membrane. This is nothing but a partition coefficient defining solute concentration between the permeate stream and at the membrane surface. All the pore transport characterizations are clubbed in this parameter that can be determined from a separate set of experiments. (2) The solute transport through the membrane pores is quantified by oversimplified equations, like Kedem-Katachalsky[3] or a modified Kedem-Katachalsky equation.[3] These equations involve several parameters that are difficult to be estimated and represent an averaged behavior without many details.

2.4.3 Third-Generation Models

In this set of models, detailed modeling of transport of species through the membrane pore is carried out. These types of models are extremely useful for ion rejection by nanofiltration membranes, transport of charged and uncharged solutes by ultrafiltration, etc.[26–29] A schematic of solute transport through a membrane pore is presented in Figure 2.6.

In these models, two different transport regions are identified. In the first case, we have the solute transport in the mass transfer boundary layer just outside of the membrane surface, in the feed side. The solute balance equation within the mass transfer boundary layer is given by the film theory, Equation (2.4). The governing equation for the solvent flow through the porous membrane is presented by a modified version of Equation (2.6) as

$$v_w = \frac{\Delta P - \Delta \pi}{\mu[R_m + k_0 C_m]} \tag{2.15}$$

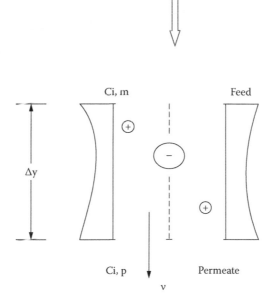

FIGURE 2.6
A schematic of solute transport through pores.

The above equation is essentially similar to Equation (2.6) with the modification of adsorption resistance of the solute in the porous membrane matrix. The adsorption resistance is assumed to be proportional to the membrane surface concentration. The solute flux through the membrane pores (assuming straight cylindrical pores) is written as, using the extended Nernst-Planck equation,[26–29]

$$j_i = -D_{i,m}\frac{dc_i}{dy} + K_{i,c}c_i v_w - \frac{z_i c_i D_{i,m} F}{RT}\frac{d\psi_m}{dy} \tag{2.16}$$

The solute and solvent flux within the membrane pore are related by

$$j_i = v_w C_{i,p} \tag{2.17}$$

Two electroneutrality conditions are used, in the bulk and within the charged membrane pores.

In the bulk of the solution,

$$I = \sum_i Fz_i j_i = 0 \tag{2.18}$$

and in the pores,

$$I = \sum_i z_i c_i + X = 0 \tag{2.19}$$

where X is the volumetric charge density within the membrane pores. The hindered coefficients for convection and diffusion in Equation (2.16) are presented as

$$K_{i,c} = \frac{v_i}{v_w} = -0.301\lambda_i + 1.022 \tag{2.20}$$

$$K_{i,d} = \frac{D_{i,m}}{D_{i,\infty}} = -1.705\lambda_i + 0.946 \tag{2.21}$$

The relation between the concentration at the membrane surface in the upstream side and at the first point within the pore is related by the Donnan partitioning. This set of equations is solved using an appropriate numerical algorithm to obtain the permeate concentration and permeate flux. The algorithm for calculations is available.[28,29]

2.5 Applications of Ultrafiltration

Ultrafiltration is becoming one of the major industrially important unit operations. It is used for various purposes, namely, separation, concentration, and fractionation. Recovery of high-valued products, recycling of permeate, and controlling the pollution can be achieved using ultrafiltration. Some of the important applications are presented below.

2.5.1 Electropaint Recovery

UF is used to control the excess conductivity in the rinse bath. UF separates water, solvents, and low molecular weight solutes from the paint and the filtrate is used to rinse the painted cars, before recycling to the feed paint tank.[3,7,30,31]

2.5.2 Textile Industry

In textiles, UF is used for the recovery of sizing chemicals. Sizing chemicals are polyvinyl alcohol, carboxymethylcellulose, etc. These are used as lubricants for the cotton before dyeing. This operation is known as sizing. Before sending the sized cotton to the dyeing step, it is thoroughly washed with water. This step is known as desizing. The desizing effluent contains a substantial amount of sizing chemicals, mentioned above. UF is used to recover these sizing chemicals.[3,32,33] Dyes are of low molecular weight, typically less than 1,000. Thus, there are examples that some of the dyes of higher molecular weights are recovered using UF membranes at lower operating pressure.

2.5.3 Metal Finishing Industry

Treatment of an oil-water emulsion is mainly done by the metal finishing industry. Grease and oil from metal pieces are removed before painting them. Thus, a chemical treatment (primary coating) is required. Before primary treatment, the metal pieces are washed with detergent solution. The rinsed water contains detergents and oily materials. UF is used to remove oil from the metal finishing rinse stream and enhances the life of the detergent bath.[3,34,35] Coolants are used to cool the machines in metal finishing industries. Thus, a huge amount of effluent is produced from the rinse water containing the oil-water mixture. UF is used to treat this effluent and controls the pollution problem. The oil-lean permeate and oil-rich retentate can be recycled back into the process. This is a step toward a zero discharge plant.

2.5.4 Dairy Industries[3,7,36,37]

Cheese whey is a supernatant stream emerging during the cheese or casein production. The volume of this liquid is as high as 90% of the original milk volume. This contains a huge amount of solids, and about 10% of that is valuable proteins. A typical concentration of solids is about 50 to 60 g/L. This solid is therefore highly nutritional. UF in the diafiltration mode can be used to concentrate these solids, and using the spray dryer, the solid food supplements are produced. UF is also used to produce protein-rich milk and is quite useful in soft cheese industries.

2.5.5 Juice Processing[38–45]

A conventional juice processing operation involves several units of operation, like enzymatic treatment, centrifugation, addition of fining agents, filtration of these fining agents by diatomaceous earth, final polish filtration, sterilization, etc. The major objectives of these processing steps are removal of polysaccharides, like pectin, high molecular weight proteins, cell debris, microorganisms, etc., to impart a high shelf life to the processed juice. These are known as haze-forming materials. The typical processing time is about 30 to 36 hours. UF can be used to remove pectin, protein, and microorganisms after pretreating fruit juice. This reduces several steps, like addition of fining agents, intermediate filtration, sterilization, etc. In fact, the process is known as cold sterilization. UF is used for processing of apples, oranges, watermelons, coconuts, kiwi fruit, bananas, etc.

2.5.6 Pulp and Paper Industry[46–50]

Effluent emerging from pulp and paper industries contains a huge amount of lignin, salt, and organic matter. Therefore, the biological (BOD) and chemical oxygen demand (COD) of these streams are extremely high and a potential threat to the environment. UF can be used to treat these streams and can address a burning pollution problem. Moreover, the pulping process requires a huge amount of inorganics, like salt and alkali. UF is used to separate lignin from black liquor produced from the kraft pulping process. Lignin is used as a good adhesive. A delignified stream can be subjected to nanofiltration to recover the salts and inorganic chemicals. Thus, the furnace used to incinerate the organics from black liquor can be replaced by UF, addressing both pollution and recovery of chemicals.

2.5.7 Tannery Industry[51-55]

A typical tannery contains several unit operations, namely, soaking, liming, deliming-bating, pickling, degreasing, tanning, and dyeing. Each of these processes consumes a huge amount of water and inorganic chemicals. Thus, the effluent emerging out of these units contains huge loads of BOD and COD. UF has been found to play an important role in treatment of supernatant generated from the pretreated effluent streams from the tanneries. UF is found to be effective in treatment of soaking, liming, degreasing, and other process streams.

2.5.8 Extraction of Costly Herbal Components from Natural Products[56-58]

Expensive components having medicinal importance are available in various products in nature. For example, polyphenols in green tea leaves, steviosides in stevia leaves, lycopenes in tomato and watermelon juice, etc. These are grouped as phytochemicals. They have antioxidant, anticarcinogenic properties. Therefore, phytochemicals have huge applications in food industries as food supplements and additives, and in pharmaceutical and cosmetic industries. Extraction of polyphenols, lycopenes, and stevioside from their water extract using UF is reported. Also, these processes are highly scalable.

2.5.9 Pharmaceutical Industries[59-62]

UF has tremendous application in this industrial sector to separate various enzymes from fermentation, plasma separation, separation and fractionation of protein solution, separation of antibodies, and to produce pyrogen-free solution. However, enzyme production from fermentation broth is the major application of UF. Fractionation of various proteins from human plasma is carried out using UF. Good quality water is a major concern in this industry. UF offers an economical alternative to process tap or distilled water by removing suspended particles, colloidal silica or colloidal iron, spores, and microorganisms.

2.5.10 Pure Water Production[63,64]

Purer water is required for many applications, e.g., boiler feed water to prevent scaling, rinsing of electronic components, beverage production, etc. Removing colloidal particles, microorganisms, suspended materials, etc., UF can be used for these purposes. It is used as a pretreatment step for reverse osmosis and ion exchange processes to avoid fouling of the process equipment.

2.5.11 Upcoming Applications

Some of the latest applications of UF include polyelectrolyte-enhanced UF (PEUF),[65,66] micellar-enhanced UF (MEUF),[67-70] electric field-enhanced UF (EUF),[70-75] etc. In PEUF the idea is that if a stream contains low molecular weight inorganic solutes (typically carcinogenic heavy metal cations), using a polyelectrolyte the metal ions can be bound to the charged locations of the polymer. Metal ion-bound polyelectrolytes, being larger in size, can be removed by UF membranes easily. Reports of removal of cadmium and other heavy metals using PEUF are available. In MEUF, by selecting a suitable surfactant, one can generate the surfactant micelles that have a hydrophobic core and hydrophilic outer surface (either positive or negative). Trace amounts of organic solutes, like benzene, aniline, cresols, phenol, etc., can be solubilized within the hydrophobic core of surfactant micelles, and oppositely charged ionic pollutants (metal ions like, cadmium, lead, zinc, etc., and anions like cyanide, chromate, permanganate, etc.) are attached on the outer surface of micelles by electrostatic attraction. In EUF, an electric field of suitable polarity is applied to remove the charged solutes from the membrane surface. This enormously reduces the deposition of solutes on the membrane surface and increases the throughput significantly. EUF is applied to increase the productivity of the filtration of protein solution, fruit juice, enzymes, etc.

References

1. Wankat, P.C. 2007. *Separation process engineering*. New York: Prentice Hall.
2. Smith, J.M., Van Ness, H.C., and Abbott, M.M. 2005. *Introduction to chemical engineering thermodynamics*. New York: McGraw Hill.
3. Bungay, P.M., Lonsdale, H.K.M., and de Pinho, N. 1986. *Synthetic membranes: Science, engineering and applications*. Dordrecht: NATO Scientific Affairs Division, D. Reidel.
4. Cheryan, M. 1998. *Ultrafiltration and microfiltration handbook*. Lancaster, PA: Technomic Publishing Company.
5. Porter, M.C. 2005. *Handbook of industrial membrane technology*. New Delhi: Crest Publishing House.
6. De, S., and Bhattacharya, P.K. 1997. Modeling of ultrafiltration process for a two component aqueous solution of low and high (gel-forming) molecular weight solutes. *J. Membr. Sci.* 136: 57–69.
7. Ho, W.S.W., and Sirkar, K.K. 1992. *Membrane handbook*. New York: Chapman & Hall.
8. Rautenbach, R., and Albrecht, R. 1986. *Membrane processes*. New York: John Wiley & Sons.
9. Blatt, W.F., Dravid, A., Michaels, A.S., and Nelsen, L. 1970. Solute polarization and cake formation in membrane ultrafiltration: Causes, consequences, and control techniques. In *Membrane science and technology*, ed. J.E. Flinn, 47–94. New York: Plenum Press.

10. Gekas, V., and Hallstrom, B. 1987. Mass transfer in the membrane concentration polarization layer under turbulent cross flow. I. Critical literature review and adaptation of existing Sherwood correlations to membrane operations. *J. Membr. Sci.* 80: 153–170.

11. van den Berg, G.B., Racz, I.G., and Smolders, C.A. 1989. Mass transfer coefficients in cross flow ultrafiltration. *J. Membr. Sci.* 47: 25–51.

12. Nakao, S., Wijmans, J.G., and Smolders, C.A. 1986. Resistance to the permeate flux in unstirred ultrafiltration of dissolved macromolecular solutions. *J. Membr. Sci.* 26: 165–178.

13. Zydney, A.L. 1997. Stagnant film model for concentration polarization in membrane systems. *J. Membr. Sci.* 130: 275–281.

14. Bhattacharya, S., and Hwang, S.T. 1997. Concentration polarization, separation factor, and Peclet number in membrane processes. *J. Membr. Sci.* 132: 73–90.

15. Aimar, P., Howell, J.A., Clifton, M.J., and Sanchez, V. 1991. Concentration polarisation build-up in hollow fibers: A method of measurement and its modelling in ultrafiltration. *J. Membr. Sci.* 59: 81–99.

16. Kleinstreuer, C., and Paller, M.S. 1983. Laminar dilute suspension flows in plate-and-frame ultrafiltration units. *AIChE J.* 29: 529–533.

17. De, S., and Bhattacharya, P.K. 1997. Prediction of mass transfer coefficient with suction in the application of reverse osmosis and ultrafiltration. *J. Membr. Sci.* 128: 119–131.

18. De, S., Bhattacharya, S., Sharma, A., and Bhattacharya, P.K. 1997. Generalized integral and similarity solutions for concentration profiles for osmotic pressure controlled ultrafiltration. *J. Membr. Sci.* 130: 99–121.

19. Probstein, R.F., Shen, J.S., and Leung, W.F. 1977. Ultrafiltration of macromolecular solutions at high polarization in laminar channel flow. *Desalination* 24: 1–16.

20. Bhattacharya, S., DasGupta, S., and De, S. 2001. Effect of solution property variations in cross flow ultrafiltration: A generalized integral approach. *Sep. Purif. Technol.* 24: 559–571.

21. Prabhakar, R., DasGupta, S., and De, S. 2000. Simultaneous prediction of flux and retention for osmotic pressure controlled turbulent cross flow ultrafiltration. *Sep. Purif. Technol.* 18: 13–24.

22. De, S., and Bhattacharya, P.K. 1996. Flux prediction of kraft black liquor in cross flow ultrafiltration using low and high rejecting membranes. *J. Membr. Sci.* 109: 109–123.

23. Minnikanti, V.S., DasGupta, S., and De, S. 1999. Prediction of mass transfer coefficient with suction for turbulent flow in cross flow ultrafiltration. *J. Membr. Sci.* 157: 227–239.

24. Ranjan, R., DasGupta, S., and De, S. 2004. Mass transfer coefficient with suction for turbulent non-Newtonian flow in application to membrane separations. *J. Food Eng.* 65: 533–541.

25. Ranjan, R., DasGupta, S., and De, S. 2004. Mass transfer coefficient with suction for laminar non-Newtonian flow in application to membrane separations. *J. Food Eng.* 64: 53–61.

26. Bhattacharjee, S., Chen, J.C., and Elimelech, M. 2001. Coupled model of concentration polarization and pore transport in crossflow nanofiltration. *AIChE J.* 47: 2733–2745.

27. Garcia-Aleman, J., and Dickson, J.M. 2004. Mathematical modeling of nanofiltration membranes with mixed electrolyte solutions. *J. Membr. Sci.* 235: 1–13.

28. Banerjee, P., and De, S. 2010. Coupled concentration polarization and pore flow modeling of nanofiltration of an industrial textile effluent. *Sep. Purif. Technol.* 73: 355–362.
29. Banerjee, P., and De, S. 2011. Modeling of nanofiltration of dye using a coupled concentration polarization and pore flow model. *Sep. Sci. Technol.* 46: 561–570.
30. Luque, S., Gómez, D., and Álvarez, J.R. 2008. Industrial applications of porous ceramic membranes (pressure driven processes). *Membr. Sci. Technol.* 13: 177–216.
31. Bjerke, B. 1980. Membrane technology and costs; state of the art. *Desalination* 35: 375–382.
32. ElDefrawy, N.M.H., and Shaalan, H.F. 2007. Integrated membrane solutions for green textile industries. *Desalination* 204: 241–254.
33. Hao, J., and Zhao, Q. 1994. The development of membrane technology for wastewater treatment in the textile industry in China. *Desalination* 98: 353–360.
34. Qin, J., Wai, M.N., Oo, M.H., and Lee, H. 2004. A pilot study for reclamation of a combined rinse from a nickel-plating operation using a dual-membrane UF/RO process. *Desalination* 161: 155–167.
35. Hesampour, M., Krzyzaniak, A., and Nyström, M. 2008. Treatment of waste water from metal working by ultrafiltration, considering the effects of operating conditions. *Desalination* 222: 212–221.
36. Tang, X., Flint, S.H., Bennett, R.J., and Brooks, J.D. 2010. The efficacy of different cleaners and sanitisers in cleaning biofilms on UF membranes used in the dairy industry. *J. Membr. Sci.* 352: 71–75.
37. Brans, G., Schroën, C.G.P.H., van der Sman, R.G.M., and Boom, R.M. 2010. Membrane fractionation of milk: State of the art and challenges. *J. Membr. Sci.* 352: 71–75.
38. Chayya, Rai, P., Majumdar, G.C., De, S., and DasGupta, S. 2010. Mechanism of permeate flux decline during microfiltration of water melon (*Citrullus lanatus*) juice. *Food Bioproc. Technol.* 3: 545–553.
39. Rai, P., and De, S. 2009. Clarification of pectin containing juice using ultrafiltration, *Curr. Sci.* 96: 1361–1371.
40. Chhaya, Rai, P., Majumdar, G.C., DasGupta, S., and De, S. 2008. Clarification of watermelon (*Citrullus lanatus*) juice by microfiltration. *J. Food Proc. Eng.* 31: 768–782.
41. Rai, P., Majumdar, G.C., Sharma, G., DasGupta, S., and De, S. 2006. Effects of various cutoff membranes on permeate flux and quality during filtration of mosambi (*Citrus sinensis* (L.) Osbeck) juice. *Food Bioprod. Proc.* 84: 213–219.
42. Rai, P., Majumdar, G.C., DasGupta, S., and De, S. 2007. Effect of various pretreatment methods on permeate flux and quality during ultrafiltration of mosambi juice. *J. Food Eng.* 78: 561–568.
43. Conidi, C., Cassano, A., and Drioli, E. 2011. A membrane-based study for the recovery of polyphenols from bergamot juice. *J. Membr. Sci.* 375: 182–190.
44. de Barros, S.T.D., Andrade, C.M.G., Mendes, E.S., and Peres, L. 2003. Study of fouling mechanism in pineapple juice clarification by ultrafiltration. *J. Membr. Sci.* 215: 213–224.
45. Onsekizoglu, P., Bahceci, K.S., and Acar, M.J. 2010. Clarification and the concentration of apple juice using membrane processes: A comparative quality assessment. *J. Membr. Sci.* 352: 160–165.
46. De, S., and Bhattacharya, P.K. 1996. Flux prediction of kraft black liquor in cross flow ultrafiltration using low and high rejecting membranes. *J. Membr. Sci.* 109: 109–123.

47. De, S., and Bhattacharya, P.K. 1996. Recovery of water with inorganic chemicals from kraft black liquor using membrane separation processes. *Tappi J.* 79: 103–111.
48. Maartens, A., Jacobs, E.P., and Swart, P. 2002. UF of pulp and paper effluent: Membrane fouling-prevention and cleaning. *J. Membr. Sci.* 209: 81–92.
49. Dal-Cin, M.M., McLellan, F., Striez, C.N., Tam, C.M., Tweddle, T.A., and Kumar, A. 1996. Membrane performance with a pulp mill effluent: Relative contributions of fouling mechanisms. *J. Membr. Sci.* 120: 273–285.
50. Puro, L., Kallioinen, M., Mänttäri, M., and Nyström, M. 2011. Evaluation of behavior and fouling potential of wood extractives in ultrafiltration of pulp and paper mill process water. *J. Membr. Sci.* 368: 150–158.
51. Cassano, A., Molinari, R., Romano, M., and Drioli, E. 2001. Treatment of aqueous effluents of the leather industry by membrane processes: A review. *J. Membr. Sci.* 181: 111–126.
52. Purkait, M.K., Bhattacharya, P.K., and De, S. 2005. Membrane filtration of leather plant effluent: Flux decline mechanism. *J. Membr. Sci.* 258: 85–96.
53. Das, C., DasGupta, S., and De, S. 2008. Steady state modeling for membrane separation of pretreated soaking effluent under cross flow mode. *Environ. Prog.* 27: 346–352.
54. Das, C., and De, S. 2009. Steady state modeling for membrane separation of pretreated liming effluent under cross flow mode. *J. Membr. Sci.* 338: 175–181.
55. Prabhavathy, C., and De, S. 2010. Treatment of fat liquoring effluent from a tannery using membrane separation process: Experimental and modeling. *J. Hazard. Mater.* 176: 434–443.
56. Todisco, S., Tallarico, P., and Gupta, B.B. 2002. Mass transfer and polyphenols retention in the clarification of black tea with ceramic membranes. *Innov. Food Sci. Emerg. Technol.* 3: 255–262.
57. Kumar, A., Thakur, B.K., and De, S. 2011. Selective extraction of (–) epigallocatechin gallate from green tea leaves using two stage infusion coupled with membrane separation. *Food Bioproc. Technol.*, doi: 10.1007/s11947-011-0580-0.
58. Chhaya, P. 2011. Clarification of mosambi (*Citrus sinensis* (L.) Osbeck) juice using membrane technology. PhD diss. (submitted), Indian Institute of Technology, Kharagpur.
59. Yoon, Y., Westerhoff, P., Snyder, S.A., and Wert, E.C. 2006. Nanofiltration and ultrafiltration of endocrine disrupting compounds, pharmaceuticals and personal care products. *J. Membr. Sci.* 270: 88–100.
60. Ko, M.K., Pellegrino, J.J., Nassimbene, R., and Marko, P. 1993. Characterization of the adsorption-fouling layer using globular proteins on ultrafiltration membranes. *J. Membr. Sci.* 76: 101–120.
61. Ölçeroğlu, A.H., Çalık, P., and Yilmaz, L. 2008. Development of enhanced ultrafiltration methodologies for the resolution of racemic benzoin. *J. Membr. Sci.* 322: 446–452.
62. Latulippe, D.R., Ager, K., and Zydney, A.L. 2007. Flux-dependent transmission of supercoiled plasmid DNA through ultrafiltration membranes. *J. Membr. Sci.* 294: 169–177.
63. Shinde, M.H., Kulkarni, S.S., Musale, D.A., and Joshi, S.G. 1999. Improvement of the water purification capability of poly(acrylonitrile) ultrafiltration membranes. *J. Membr. Sci.* 162: 9–22.

64. M'Bareck, C.O., Nguyen, Q.T., Alexandre, S., and Zimmerlin, I. 2006. Fabrication of ion-exchange ultrafiltration membranes for water treatment. I. Semi-interpenetrating polymer networks of polysulfone and poly(acrylic acid). *J. Membr. Sci.* 278: 10–18.

65. Tabatabai, A., Scamehorn, J.F., and Christian, S.D. 1995. Economic feasibility study of polyelectrolyte-enhanced ultrafiltration (PEUF) for water softening. *J. Membr. Sci.* 100: 193–207.

66. Juang, R.S., and Chen, M.N. 1996. Retention of copper(II)—EDTA chelates from dilute aqueous solutions by a polyelectrolyte-enhanced ultrafiltration process. *J. Membr. Sci.* 119: 25–37.

67. Zeng, G.M., Xu, K., Huang, J.H., Li, X., Fang, Y.Y., and Qu, Y.H. 2008. Micellar enhanced ultrafiltration of phenol in synthetic wastewater using polysulfone spiral membrane. *J. Membr. Sci.* 310: 149–160.

68. Huang, J.H., Zhou, C.F., Zeng, G.M., et al. 2010. Micellar-enhanced ultrafiltration of methylene blue from dye wastewater via a polysulfone hollow fiber membrane. *J. Membr. Sci.* 365: 138–144.

69. Kim, C.K., Kim, S.S, Kim, D.W, Lim, J.C, and Kim, J.J. 1998. Removal of aromatic compounds in the aqueous solution via micellar enhanced ultrafiltration. Part 1. Behavior of nonionic surfactants. *J. Membr. Sci.* 147: 13–22.

70. Huang, J., Zeng, G.M., Fang, Y.Y., Qu, Y.H., and Li, X. 2009. Removal of cadmium ions using micellar-enhanced ultrafiltration with mixed anionic-nonionic surfactants. *J. Membr. Sci.* 326: 303–309.

71. Sarkar, B., and De, S. 2011. Prediction of permeate flux for turbulent flow in cross flow electric field assisted ultrafiltration. *J. Membr. Sci.* 369: 77–87.

72. Sarkar, B., and De, S. 2010. Electric field enhanced gel controlled cross-flow ultrafiltration under turbulent flow conditions. *Sep. Purif. Technol.* 74: 73–82.

73. Sarkar, B., DasGupta, S., and De, S. 2009. Electric field enhanced fractionation of protein mixture using ultrafiltration. *J. Membr. Sci.* 341: 11–20.

74. Sarkar, B., De, S., and DasGupta, S. 2008. Pulsed-electric field enhanced ultrafiltration of synthetic and fruit juice. *Sep. Purif. Technol.* 63: 582–591.

75. Sarkar, B., DasGupta, S., and De, S. 2008. Cross-flow electro-ultrafiltration of mosambi (*Citrus sinensis* (L.) Osbeck) juice. *J. Food Eng.* 89: 241–245.

3

Surfactants

A surfactant is popularly visualized as a surface-active agent that alters the surface properties of a solution, resulting in its multifarious effects and benefits. Surfactants are added to a solution to change its surface tension by minimizing the effective free surface energy of the interfaces. The adsorption of a surfactant molecule is driven by lowering of the interfacial tension in the phase boundary. The degree of surfactant adsorption at the interface depends on surfactant structure and the nature of the two phases that meet the interface.[1] A surfactant molecule is typically a long hydrocarbon chain or an aromatic ring with a functional group acting as its head. Typically, the tail is derived from an unsaturated high molecular weight fatty acid or its derivatives. The presence of an electrophilic or nucleophilic group is responsible for its hydrophilic nature; consequently, the tail acts as a hydrophobic end of the molecule. The hydrocarbon chain interacts weakly with the water molecules in an aqueous environment, whereas the polar or ionic head group interacts strongly with water molecules via dipole or ion-dipole interactions. It is this strong interaction with the water molecules that renders the surfactant soluble in water. However, the cooperative action of dispersion and hydrogen bonding between the water molecules tends to squeeze the hydrocarbon chain out of the water, and hence these chains are referred to as hydrophobic. For a compound to be a surfactant, it should possess three characteristics: the molecular structure should be composed of polar and nonpolar groups, it should exhibit surface activity, and it should form self-assembled aggregates (micelles, vesicles, liquid crystalline, etc.) in liquids.

3.1 Types of Surfactants

The surfactant molecules are amphibilic in nature, since they contain both the hydrophilic and hydrophobic groups. In general, the surfactants are categorized mainly as anionic, cationic, nonionic, and zwitterionic. Figure 3.1 shows an outline of the classification of surfactants.

3.1.1 Anionic Surfactant

These are the most widely used category of surfactants[2] because of their low cost of manufacture and wide variety of application. Due to the presence of

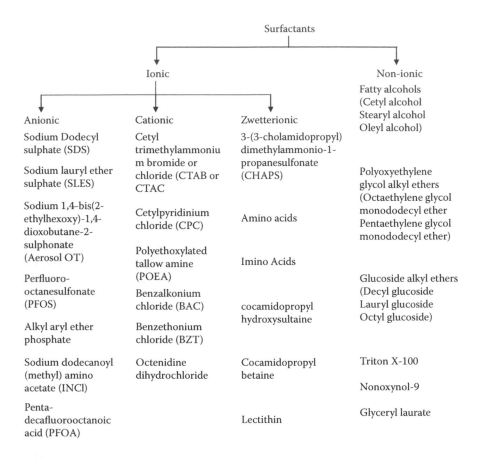

FIGURE 3.1
Generalized classification of surfactants based on their ionic properties.

nucleophilic groups, anionic surfactants ((SDS)), when put in aqueous solution, produce negatively charged heads. Because of their negative charge, the surfactants are sensitive to pH and electrolyte concentration. These surfactants dissociate in water as mono-, di-, or trivalent metal ions such as carboxylates dissociate as $RCOO^-$ and M^+, sulfates dissociate as RSO_4^- and M^+, sulfonates dissociate as RSO_3^- and M^+, and phosphates dissociate as $ROP(O)O_2^{2-}$ and M_2^+, where R is the hydrophobe and M is the organic or inorganic counterion. The hydrophobic chain is a linear combination of alkyl groups of 12 to 16 carbon atoms. Linear chains are preferred since they are more effective and more easily degraded than branched ones.

3.1.1.1 Carboxylates

Most commercial soaps are under this category of surfactants, for example, sodium or potassium stearate, $C_{17}H_{35}COONa$, and sodium myristate,

$C_{14}H_{29}COONa$. The soaps are a mixture of fatty acids obtained from tallow, coconut oil, palm oil, etc. The main attraction of these simple soaps is their low cost, ready biodegradability, and low toxicity. Their main disadvantages are their ready precipitation in water containing bivalent ions such as Ca^{2+} and Mg^{2+}, which are present commonly in hard water. To avoid such precipitation in hard water, the carboxylates are modified by introducing some hydrophilic chains, e.g., ethoxy carboxylates with the general structure $RO(CH_2CH_2O)_nCH_2COO^-$, ester carboxylates containing hydroxyl or multi-COOH groups, and sarcosinates that contain an amide group with the general structure $RCON(R')COO^-$.

Addition of the ethoxylated groups increases water solubility and enhances chemical stability (no hydrolysis). The modified ether carboxylates are also more compatible, both with electrolytes and with other nonionic, zwitterionic, and sometimes even cationic surfactants. Phosphate esters have interesting properties, being intermediate between ethoxylated nonionics and sulfated derivatives. They have good compatibility with inorganic builders and can be good emulsifiers.[3]

3.1.1.2 Sulfates

These are the most significant class of surfactants produced by the reaction of an alcohol and sulfuric acid. Sulfur dioxide or chlorosulfonic acid is commonly used for sulfating alcohols. The properties of sulfate surfactants depend on the nature of the alkyl chain and the sulfate group. The alkali metal salts show good solubility in water, but tend to be affected by the presence of electrolytes and acidity/alkalinity of the solution. The most common sulfate surfactant is sodium dodecyl sulfate (SDS; sometimes referred to as sodium lauryl sulfate), which is extensively used both for fundamental studies and in many industrial applications. At room temperature (298°K), this surfactant is quite soluble and 30% aqueous solutions are fairly fluid (low viscosity). The sulfate esters have gained a lot of attention due to their good water solubility and surface activity, as well as reasonable chemical stability, a relatively simple synthetic pathway amenable to low-cost commercial production, and readily available starting materials from a number of sources.

Like carboxylates, the sulfates are chemically modified to enhance their surface properties. Introduction of some ethylene oxide units in the chain improves solubility in contrast to straight chain alkyl sulfates. When a fatty alcohol (ROH) is ethoxylated, the resulting ether still has a terminal –OH, which can subsequently be sulfonated to give the alcohol ether sulfate $R(OCH_2CH_2)_nOSO_3^-M^+$. Also, the ether sulfates are chemically more stable and compatible in the presence of electrolytes in the solution. For example, sodium dodecyl 3-mole ether sulfate, which is essentially dodecyl alcohol, reacted with 3 moles ethylene oxide (EO) and then was sulfated and neutralized by NaOH. The critical micelle concentration of the ether sulfates is much lower than that of the surfactant without ethoxylation.

3.1.1.3 Sulfonates

Unlike the sulfates, the sulfur atom is directly linked to the carbon atom attached to the alkyl group in the sulfonates. Alkyl aryl sulfonates are the most common type of these surfactants (e.g., sodium alkyl benzene sulfonate), and these are usually prepared by reaction of sulfuric acid with alkyl aryl hydrocarbons, e.g., dodecyl benzene. Chemical modification is done by introducing ethylene oxide units into the surfactants example, e.g., sodium nonyl phenol 2-mole ethoxylate ethane sulfonate ($C_9H_{19}C_6H_4(OCH_2CH_2)_2SO_3\ Na^+$).

Chlorosulfonic acid (CSA), HSO_3Cl, has also been used as an effective sulfonating agent. The effectiveness of chloride as a leaving group and the absence of water as a by-product means that chlorosulfonation can be run at stoichiometry close to 1:1 and with efficient conversion of the organic substrate. The most common and cost-effective sulfonating agent in use is sulfur trioxide itself, which has the benefit of being a highly aggressive sulfonating agent, with no direct by-products of sulfonation. The surface activity of this type of surfactants is not appreciable until the number of carbon atoms is 8, in the case of alkyl sulfonates, and is a problem in solubility when it reaches 18. The sulfonates are more stable to hydrolysis than the analogous alkyl sulfates. Linear alkyl benzene sulfonates (LABSs) are manufactured from alkyl benzene, and the alkyl chain length can vary from C_8 to C_{15}; their properties are mainly influenced by the average molecular weight and the spread of the carbon number of the alkyl side chain. It has been suggested that the paraffin sulfonates have high water solubility and lower viscosity than the LABSs of comparable chain length.[3] Sulfonation of benzene and naphthalene typically constitutes the class of alkylaryl sulfonates. Simple sulfonated aromatic groups alone are not able to impart sufficient surface activity, but when the aromatic ring is substituted with one or more alkyl groups, the surface-active character of the molecule is greatly enhanced. These are not seriously affected by the presence of calcium or magnesium salts or high concentration of other electrolytes, and they are also fairly resistant to hydrolysis in the presence of strong mineral acids or alkali.

A special class of sulfonates is the sulfosuccinates ($R_1O_2CCH_2CH(SO_3^-M^+)$ CO_2R_2) and sulfosuccinate esters and diesters, which are esters of sulfosuccinic acid (The commercial name of sulfosuccinate esters is aerosols). These materials exhibit excellent wetting, emulsifying, dispersing, and foaming properties, are relatively cheap, and are easy to prepare, which has given them wide acceptance. Their major disadvantage is that the ester group is susceptible to hydrolysis in either alkaline or acidic solutions.[4]

3.1.1.4 Phosphates

The phosphoric acid esters are a special group of anionic surfactants. These surfactants are somewhat superior than the sulfates and sulfonates in some applications because of their foaming characteristics, good solubility in water and many organic solvents, and resistance to alkaline hydrolysis. But the phosphates have a disadvantage of being economically more expensive than the sulfates and sulfonates. Useful modifications of the alkyl phosphate surfactants are those in which a polyoxyethylene chain is inserted between the alkyl and phosphate ester groups. An advantage of the addition of the phosphate group to the nonionic is that the resultant material will often have better solubility in aqueous electrolyte solutions. Also, the pour point, the temperature at which the nonionic surfactant solidifies, can be significantly decreased by phosphorylation. These materials are somewhat limited by the fact that they hydrolyze in the presence of strong acids, although their stability to strong alkali, like that of the simple phosphates, is quite good. The wetting, emulsifying, and detergent properties of these surfactants are generally better than those of similar phosphate surfactants not containing the added solubilizing groups.[3]

3.1.2 Cationic Surfactant

The cationic surfactants (cetyl pyridinium chloride [CPC]), when put in aqueous medium, produce positively charged heads due to the presence of electrophilic groups. Cationic surfactants are sensitive to pH and electrolyte content of the solution. Cationic surfactants show strong adsorption to solid surfaces. The hydrophobe may be attached directly or indirectly to a quaternary ammonium group, a protonated amino group, or a heterocyclic base.[4] The most widely used cationic surfactant is the quaternary ammonium compounds[5] with the general structure shown in Figure 3.2. In the figure, R represents the organic group and X represents the halide group. The organic group can be an alkyl group, aryl group, or alkyl-aryl group. The cationic surfactant is generally water soluble when there is only one long alkyl group. They are generally compatible with most inorganic ions and hard water, but they are incompatible with metasilicates and highly condensed phosphates. They are incompatible with protein-like materials and anionic surfactants. They are generally stable to the acidic and alkaline environments.

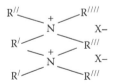

FIGURE 3.2
General structure of a cationic surfactant.

3.1.3 Zwitterionic Surfactant

These are the class of surfactants that contain both cationic and anionic groups.[6] The characteristic phenomenon is the dependence of their properties on the pH of the solution. In acidic solution (pH < 7), the surfactant molecule acquires a positive charge, while in alkaline solution (pH > 7), the molecule acquires primarily a negative charge. But, there exists a specific pH at which the molecule has equal positive and negative groups (isoelectric point). At this point, the surfactant behaves with both an anionic and a cationic nature and the hydrophilic part is internally neutralized by the positive and negative counterions. The zwitterionics, due to their ability to support both positive and negative charges, usually have large head groups, the hydrophilic portion of the molecule that exhibits an affinity for the aqueous phase. The zwitterionics are generally used together with anionic or nonionic surfactants to modify the solubility,[7] micelle size,[8] foam stability,[9] detergency, and viscosity.[10] The surface activity of zwitterionic surfactants widely varies depending upon the relative distance between the opposite charges, having maximum activity at the isoelectric point. The characteristics of these materials closely resemble the properties of the nonionics at the isoelectric point. However, there are some materials that are zwitterionic at all pH values (these materials do not have any isoelectric point), for example, sulfobetaines, sultaines, etc.[10]

3.1.4 Nonionic Surfactant

The electrically neutral property of a surfactant is advantageous, particularly due to reduced sensitivity of interaction of the molecule in a solution containing various other charged species. This property of the nonionics makes this class of surfactant universally compatible in any solution. Most of the surfactants belong to the polyoxyethylene family, commonly referred to as the ethoxylated surfactants.[11,12] An interesting property of this group of surfactants is the inverse temperature relationship; i.e., as the solution temperature is increased, the solubility of the surfactant in water is decreased.

The nonionics can be made resistant to hard water, polyvalent metallic ions, and strong electrolytic solution. However, the limitations of these classes of surfactants are the products being liquids or pastes (rarely solid pellets) and decreased solubility on heating (inverse temperature relationship).[13]

The polyoxyethylene (POE)-based surfactants have the general formula $RX(CH_2CH_2O)_nH$, where R is the typically surfactant hydrophobic group, but may also be a hydrophobic polyether such as polyoxypropylene, and X is the oxygen atom (O) or nitrogen atom (N), or another functionality capable of linking the POE chain to the hydrophobe, and n is the number of oxyethylene units in the hydrophilic group. In general, the properties of materials change with varying the POE chain lengths of the same hydrophobic group,[14,15] for example, in the case of the n-dodecyl-POE surfactant. Following are the typical features of this group of surfactants:

- Water solubility increases as the number of the oxyethylene groups increases from 3 to 16.
- The surface tension of aqueous solutions of the materials decreases over the same composition range.
- The interfacial tension between aqueous solutions and hydrocarbons reaches a maximum at around $n \sim 5$ and decreases from there.

Foaming power reaches a maximum at about $n = 5$ or 6 and remains relatively constant from that point.

3.2 Structure

Typically a surfactant molecule has a hydrophobic part that is referred to as the tail and a hydrophilic part that is referred to as the head, as shown in Figure 3.3.

The hydrophobic part of a surfactant may be branched or linear. The polar head groups are usually, but not always, attached to one end of the alkyl chain. The degree of chain branching, the position of the polar group, and the length of the chain are parameters of importance for the physicochemical properties of the surfactant. An increase in the length of the hydrophobic group leads to the following consequences: (1) the surfactant solubility in water decreases and in organic solvents increases, (2) the packing of surfactant molecules at the interface becomes dense, (3) the tendency of the surfactant to adsorb at an interface or to form micelles increases, (4) the melting point of surfactants increases, and (5) the probability of formation of liquid crystal phases increases. The presence of counterions beyond a particular concentration may lead to precipitation of ionic surfactant in aqueous solution.[11] The relative size of the hydrophobic and polar groups, not the absolute size of either of the two, is decisive in determining the physicochemical behavior of a surfactant in water.

The introduction of branching or unsaturation into the hydrophobic group leads to (1) an increase in the solubility of the surfactant in water or in organic

Head
(Hydrophilic)

Tail
(Hydrophobic)

FIGURE 3.3
A schematic of a typical surfactant molecule.

solvents (compared to the straight-chain, saturated isomer); (2) decreases in the melting point of the surfactant and of the adsorbed film; (3) enhancement of looser packing of the surfactant molecules at the interface (the cis isomer is particularly loosely packed; the trans isomer is packed almost as closely as the saturated isomer); (4) inhibition of liquid crystal phase formation in solution; (5) an increase in oxidation and color formation in unsaturated compounds; (6) a decrease in biodegradability in branched-chain compounds; and (7) an increase in thermal instability.[16]

The presence of an aromatic nucleus in the hydrophobic group increases the adsorption of the surfactant onto polar surfaces, decreases its biodegradability, and causes looser packing of the surfactant molecules at the interface. Polyoxypropylene units increase the hydrophobic nature of the surfactant, its adsorption onto polar surfaces, and its solubility in organic solvents. Polyoxyethylene units decrease the hydrophobic character of the surfactant. The presence of either of these groups as the hydrophobic group in the surfactant permits reduction of the surface tension of water to lower values than those attainable with a hydrocarbon-based hydrophobic group. Perfluoroalkyl surfaces are both water and hydrocarbon repellent.[17]

3.3 Properties

Surfactants are surface-active reagents that reduce the surface tension and exhibit a tendency to form aggregates in solvents. The mode of action is directly related to the chemical structure. All surfactants have in common an asymmetric skeleton with a hydrophobic core and a hydrophilic moiety. Due to this bifunctionality, both parts of the surfactant molecule interact differently with water, the most commonly used solvent with surfactants. Surfactant adsorption is a consideration in any application where surfactants come in contact with a surface or interface. Adsorption of surfactants may lead to positive effects, as in surface wettability alteration, or be detrimental, as in the loss of surfactants from solution. When surfactant molecules adsorb at an interface, they provide an expanding force and cause the interfacial tension to decrease (at least up to the CMC). This is illustrated by the general Gibbs adsorption equation, from which the packing density of the surfactant in a monolayer and the area per adsorbed molecule can be calculated. If the interface undergoes a sudden expansion, a surface tension gradient is established that induces liquid flow in the near-surface region, termed the Marangoni effect. The hydrophobic group is generally surrounded by water molecules, which results in good solubility. On the other hand, the hydrophobic moieties are repulsed by strong interactions between the water molecules.[18] Hence, the molecules are rejected out of the inner phase and accumulate at the interfaces. Surfactant adsorption may cause a surface electric charge to increase, decrease, or not significantly change at all. For example, an oil-aqueous interface can become negatively charged in alkaline aqueous solutions due to the ionization of surface carboxylic acid groups, the adsorption of natural surfactants present in the oil, and the adsorption of charged mineral solids.[19,20] The presence of a surface charge influences the distribution of nearby ions in a polar medium and an electric double layer (EDL) is formed, consisting of the charged surface and a neutralizing excess of counterions over co-ions, distributed near the surface. Most colloidal dispersions, including emulsions, suspensions, and foams, are not thermodynamically stable, but may possess some degree of kinetic stability. The stability of the dispersion depends upon how the particles interact when this happens.[21] Surfactant adsorption at liquid interfaces can influence emulsion stability by lowering interfacial tension, increasing surface elasticity, increasing electric double-layer repulsion (ionic surfactants), lowering the effective Hamaker constant, and possibly increasing surface viscosity. The nature of the surfactant can determine the arrangement of the phases in an emulsion, namely, the dispersed and continuous phases.

3.4 Formation of Micelle

The adsorption of a surfactant to the solid-liquid or liquid-liquid interfaces leads to the formation of the aggregates. The self-association of amphiphilic molecules or ions in aqueous solution to form micelles is governed by two opposing forces. The hydrophobic force favors expulsion of the hydrophobic tail of the amphiphile from the aqueous medium with formation of a fluid organic droplet that constitutes the micelle core. The polar head groups extend into the aqueous medium from this core, and the force of repulsion between them limits self-association to relatively small aggregates. Formation of micelles of size m from monomeric amphiphile (Z) is governed by a set of equilibria, $mZ \rightleftharpoons Z_m$. Assuming thermodynamic ideality, the equilibrium can be expressed in terms of a set of equilibrium constants, $K_m = [Z_m]/[Z]^m$. An entirely equivalent and more convenient expression is in terms of the unitary free energy ΔG_m^0 for transfer of a single amphiphile from aqueous solution to a micelle of size m.[22]

$$\ln X_m = -m\frac{\Delta G_m^0}{RT} + m\ln X_1 + \ln m \tag{3.1}$$

where X_1 is the mole fraction of amphiphile in monomeric form and X_m is the mole fraction incorporated in micelles of size m, i.e., $X_m = m[Z_m]$.

Surfactant aggregation in a solution alters the physical properties of the solution. The formation of clusters, popularly termed micelle, leads to a drastic change in the physicochemical properties of the solution. The onset of sharp variation of such bulk properties at a particular concentration is the critical micellar concentration that is specific to a particular surfactant.

At the regime of critical micellar concentration (CMC), the colligative properties of surfactants in solution do not vary in a simple way with concentration. At very low concentration, properties such as equivalent conductivity are similar to those of the conventional electrolytes. However, the gradient of conductivity sharply decreases beyond the CMC. Similarly, almost all the properties change in that narrow range of concentration close to CMC. The nature of the change of osmotic pressure and light scattering clearly infers that large clusters are formed beyond the CMC. Intensity of light scattering at an angle 90° to an incident beam by surfactant solution depends on the concentration, as shown in Figure 3.4. Below the CMC, the scattering effects are small, consistent with molecularly dispersed surfactant. The rapid rise of solution turbidity, above the CMC, indicates the increasing concentration of much larger aggregated species. A marked decrease in the diffusion coefficient, as the surfactant concentration is increased beyond the CMC, signifies that the mobility of the micelles is less than that of an isolated surfactant molecule.[23] The most important bulk property, surface tension as a function of surfactant concentration, decreases with concentration and becomes constant beyond CMC.

Since the presence of the surfactant molecules decreases the available free interface, the surface energy decreases almost linearly with increasing concentration. Added surfactant molecules prefer to move to the surface to

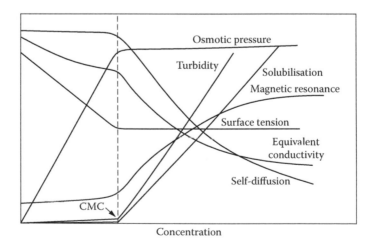

FIGURE 3.4
Variation of the physicochemical properties over a wide range of concentrations.

minimize the overall energy of the system due to their hydrophilic align-
ment of the head toward the aqueous region. However, when the available
surface area of the system is reduced significantly (at the CMC), the excess
surfactant molecules result in the formation of the micelle.[24–26] So, above the
so-called CMC, additional surfactant exists as aggregates or micelles.

When a molecule containing both a hydrophobic group and a hydrophilic
group is introduced into water, a distortion of the water structure to accom-
modate the solute molecules occurs, disrupting the accommodation of the
water molecules and requiring them to orient around the hydrophobic tail in
a more floating iceberg-like structure.[27] That more structured arrangement
increases the free energy (basically the entropy) of the system. The physical
result of such an energy increase is a tendency for the surfactant molecules
to be adsorbed at available interfaces where preferred molecular orientations
may serve to reduce the total free energy of the solution, or for the formation
of molecular aggregates with their hydrophobic portions directed toward the
interior of the micelle. Micellization, therefore, is an alternative mechanism
to adsorption for the reduction of solution free energy by the minimization
of the distortion of the structure of the bulk water. Although the removal of
the hydrophobe from the water environment results in a decrease in energy,
the adsorbed or aggregated hydrophobe may experience a loss of freedom
(decrease in entropy) that will thermodynamically reduce the attractiveness
of the process.[28]

3.4.1 Factors That Influence the CMC of the Solution

When the temperature of measurement is above the Krafft temperature
(temperature below which the surfactant is not rally soluble), the dependence of CMC on the nature of the hydrophobic group is simple. For hydrophobic groups consisting of n-alkyl chains, the addition of each CH_2 group
lowers the CMC by a constant factor so that log(CMC) varies linearly with
chain length.[29] In aqueous medium, CMC decreases as the number of carbon atoms increases up to a particular value, and thereafter the value does
not decrease considerably.[30] For nonionics and zwitterionics, the decrease
of CMC with the number of carbon atoms is somewhat larger than that
of the ionics. Structural variations such as chain branching or molecules
having more than one hydrophobic chain have less influence on the CMC.
However, the presence of unsaturation or the cis isomer in the hydrophobic chain increases the CMC compared to a saturated chain or trans
isomer. Also, introduction of a polar hydrophobic group causes a significant rise in the CMC in the aqueous solution. Surfactants having multiple
polar head groups strongly inhibit the hydrophobic effect of the methylene
groups.[31] In aqueous medium, ionic surfactants have much higher CMCs
than nonionic surfactants containing equivalent hydrophobic groups. As
the hydrophilic group is moved from a terminal position to a more central
position, the CMC increases.[32] Also, the CMC is higher when the charge

on an ionic hydrophilic group is closer to the α-carbon atom of the (alkyl) hydrophobic group.[33]

In the case of an ionic surfactant, the degree of dissociation of the polar group depends on the pH of the solution.[34] In general, the CMC will be high at pH values when the group is charged (low pH for $-NH_2$ and high pH for $-COOH$) and low when the groups are uncharged.

In the case of ionic surfactants, the CMC is reduced by adding electrolytes. On adding electrolytes, the force of electrostatic repulsion between the polar head groups is significantly decreased, thus enabling the micelles to form easily at lower concentration. For nonionic surfactants, electrolytes capable of "salting out" reduce the CMC (for example, NaCl, KCl, NaBr, and $NaNO_3$).[35] The larger the hydrated radius of the counterion, the weaker the degree of binding; thus $NH_4^+ > K^+ > Na^+ > Li^+$ and $I^- > Br^- > Cl^+$. The extent of binding of the counterion increases with the polarizability and charge of the counterion and decreases with an increase in its hydrated radius. Thus, in aqueous medium, for the anionic lauryl sulfates, the CMC decreases in the order $Li^+ > Na^+ > K^+ > Cs^+ > N(CH_3)_4^+ > N(C_2H_5)_4^+ > Ca^{2+}, Mg^{2+}$, which is the same order as the increase in the degree of binding of the cation.[36] A small amount of organic additives produces a marked change in the CMC of the solution. The CMC can be affected by incorporation of the organic materials into the micelle or by modifying the solvent-micelle interactions.

The effect of temperature on the CMC of surfactants in aqueous medium is complex, the value appearing first to decrease with temperature to some minimum and then to increase with further increase in temperature. Temperature increase causes decreased hydration of the hydrophilic group, which favors micellization. However, an increase in temperature also causes disruption of the structured water surrounding the hydrophobic group, an effect that does not favor micellization. The relative magnitude of these two opposing effects therefore determines whether the CMC increases or decreases over a particular temperature range.[37–40]

3.5 Thermodynamics of Micelle Stability

There exists a dynamic equilibrium between the monomers and micelles present in the solution. The surfactant molecules constantly orient themselves either at the free interface or in the bulk solution or in the micelle group. The exchange of the surfactant molecules is summarily outlined in Figure 3.5.

Thermodynamics of micelle formation is very well described by two simple models: mass action model[41] and phase separation model.[42] Consider a cationic micelle ($M^{(n-m)+}$), assumed to contain n detergent ions (D^+) and m firmly bound counterions (C^-), so that a fraction m/n of the charge of the

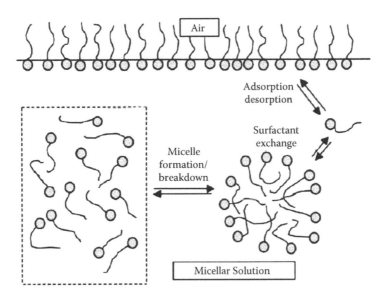

FIGURE 3.5
Schematic of the dynamics of the surfactant molecules in solution.

detergent ions in each micelle is neutralized. The process of micellization can be represented by a reversible reaction[43] as

$$nD^+ + mC^- \rightleftharpoons M^{(n-m)+}$$ (3.2)

The equilibrium constant for the above equation is

$$K_M = \frac{[M^{(n-m)+}]}{[D^+]^n[C^+]^m}$$ (3.3)

The equilibrium (Equation (3.3)) can be broken up one micelle molecule at a time. This multiple micelle equilibrium model can be expressed as

$$K_M = \prod_2^n K_Q$$ (3.4)

where K_Q is the individual stepwise association constant. It must be noticed that K_M is not the actual equilibrium constant.[44] The thermodynamic correlation based on standard free energy change of micelle formation per mole of monomer concentration can be described by

$$\Delta G_m^0 = -\frac{RT}{n} \ln K_M$$ (3.5)

The above can be more simply expressed as

$$\Delta G_m^0 = \frac{RT}{n}(n\ln[D^+] + m\ln[C^-] - \ln[M^{(n-m)+}])$$ (3.6)

In the case of a polydispersed micellar solution, ΔG_m^0 represents the free energy for the addition of a single monomer to a micelle with the most probable size.[45] The stability of the micelles is expressed by the sign of the free energy change of the solution.

On the other hand, in the case of a phase separation model, the micelles together with the bound counterions are considered a separate phase (with phase separation occurring in the zone of the CMC). By the definition of phase rule, the monomers and micelles are in equilibrium with each other only at the transition concentration (CMC). Above the CMC, the chemical potential for the monomer is practically constant. Thus, the free energy change of the micellar solution can be derived as[46]

For nonionic micelle:　　$\Delta G_p^0 = RT\ln(\text{CMC})$ (3.7)

For ionic micelle:　　$\Delta G_p^0 = RT\left(1 + \frac{x}{y}\right)\ln(\text{CMC})$ (3.8)

where y is the aggregation number and x is the number of counterions bound per micelle. In the case of total counterion binding, $x/y = 1$ and the micelle is electrically neutral.[47] It must be realized that the above expressions of both the models defines a preliminary state of the system involving a number of assumptions that include the uniformity of micelle size (monodispersed), negligible interactions between the micelles at the CMC, and the absence of any other solvent molecules other than the counterions and the surfactant ions.

The stability of a mixed micellar system (mixture of ionic and nonionic surfactant) can be analyzed similarly using a phase separation model. A solution containing n ionic species (A) and $(m - n)$ nonionic species (B) forms a mixed micelle (M) having a composition α_m. The standard state is considered the hypothetical pure state of infinite dilution in a salt solution for three species: micelles, ionic monomer, and nonionic monomer. The ionic species are completely solubilized, hence $n = \alpha_m m$. The above micellization can be expressed in a reversible rate equation:

$$m\alpha_M A + m(1 - \alpha_M)B \rightleftharpoons M$$ (3.9)

Neglecting the micelle-micelle interaction and having moderately high ionic surfactant concentration than the nonionic concentration to favor more micelles in the solution, the standard free energy of the state of the system can be expressed as[48]

$$\frac{\Delta G^0_{mm}}{RT} = \ln(\text{CMC}) + (1-\alpha_M)\ln\left(\frac{1-\alpha_1}{1-\alpha_M}\right) + \alpha_M \ln\left(\frac{\alpha_1}{\alpha_M}\right) \qquad (3.10)$$

where α_1 is the nonionic surfactant composition in the solution. It is to be mentioned here that the above expression coincides with Equation (3.7) whenever the micelle and monomer composition are similar in the solution. Theoretical analysis of multicomponent surfactant mixtures and the phenomenon of micellelization have been analyzed by Goldsipe and Blankschtein.[49] Recently, quantification of micellar free energy[50] and hydrophobic effect[51] using the molecular thermodynamic model has also been investigated, using computer-aided molecular dynamic simulation.

3.6 Micelle Characteristics

The shape of micelles has been a long debated issue. Hartley proposed that micelles are spherical in shape, with the charged groups residing on the outer surface,[52,53] while others claim the coexistence of lamellar and spherical shapes,[54] a sandwich or lamellar model,[55] only rod shaped,[56] etc. The formation of micelles by ionic surfactants is ascribed to a balance between hydrocarbon chain attraction and ionic repulsion, whereas, for nonionic surfactants, the hydrocarbon chain attraction is opposed by the requirements of hydrophilic groups for hydration and space. Thus, the micellar structure is determined by an equilibrium between the repulsive forces among hydrophilic groups and the short-range attractive forces among hydrophobic groups. The small-angle neutron scattering experiments on sodium dodecyl sulfate and other ionic micelles support the basic Hartley model of a spherical micelle.[57–59] However, as the ion concentration is increased, the shape of ionic micelles changes in the sequence spherical-cylindrical-hexagonal-lamellar.[60–63] For nonionic micelles, on the other hand, the shape seems to change from spherical directly to lamellar with increasing concentration.[64,65]

The micelles are formed by aggregation of monomer surfactant dispersed in the solvent. This phenomenon is opposed by an increase in electrostatic energy (for ionic surfactants) and a decrease in entropy due to association. These unfavorable conditions suggest that micellization is associated with an energy decrease resulting from the condensation of hydrophobic groups of surfactant molecules into a micellar aggregate. The cause of such behavior is complex ordered orientation of the water molecules. Water molecules surrounding a hydrophobic solute become more ordered than bulk water molecules and have lower entropy, with an increase in hydrogen bonding in this region. Favorable free energy for transfer of a nonpolar solute from an aqueous to a hydrophobic environment arises from a large positive entropy associated with the disordering of water molecules in the vicinity of nonpolar solutes.[66,67] The driving force for micellization results from the transfer of nonpolar surfactant chains from an ordered aqueous environment to the hydrocarbon-like environment of the micelle interior, even though a large negative entropy would otherwise have been expected for the transfer of surfactant molecules and counterions from aqueous solution to the confines of a small micelle.

The hydrophilic groups of surfactants make the otherwise hydrophobic molecule partly soluble, also restricting micelle formation, keeping the micelle size within a certain limit. This occurs because the hydrophilic groups have a great affinity for water molecules by an ion-induced dipole interaction for ionic surfactants and by hydrogen bonding for nonionic surfactants. Water molecules of an ionic group come to lack hydrogen bonds under the influence of the strong electric field and become molecules of hydration. Water molecules between these nearest neighbors and the outer bulk water are therefore the least restricted, because of the partial destruction of hydrogen bonds, induced by the central ion.

The CMC of ionic surfactants possesses a minimum with increasing temperature (refer to Figure 3.6). The apparent entropy change of micellization (which is based only on a CMC change with temperature) decreases from a positive value to a negative one with increasing temperature. The positive entropy change substantiates the idea that the hydrophobic bond supplies a motive force. The negative entropy change, on the other hand, indicates that with increasing temperature the ordinary condensation effect becomes stronger than hydrophobic effects.[68-71] The micelle aggregation number of ionic surfactants decreases almost linearly with temperature.[72-75] The CMC values for nonionic surfactants are less by one or two orders of magnitude than those of ionic surfactants with the same hydrophobic group, and decrease with temperature.[75] Specifically, the changes in entropy and enthalpy are positive, and decrease with temperature while remaining positive. The micellar aggregation number of nonionic surfactants increases rapidly with temperature.[76,77] Regarding aggregation number, there still remains a question: whether the apparent increase of the aggregation number is based on an actual increase in the aggregation number or on an altered spatial arrangement due to increasing interaction among micelles of smaller size as temperature approaches the cloud point.[78,79]

The effect of pressure on both ionic [80-84] and nonionic surfactants[85-87] is considerable at higher magnitudes. For example, the increase in CMC of SDS is maximum at 1,000 atm, by 15%. The change in partial molar volume in micelle formation is from a positive to a negative value as this pressure

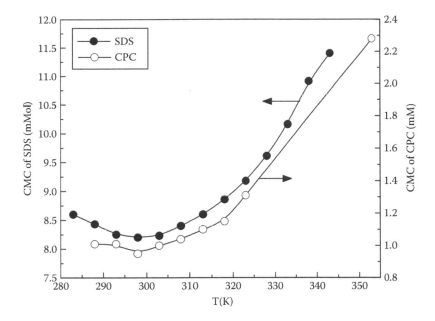

FIGURE 3.6
Variation of CMC with temperature for both anionic and cationic surfactants.

is exceeded, due to the greater compressibility of surfactant molecules in the micellar state. The partial molar volume change follows a linear relationship with the number of carbon atoms present in the surfactant molecule.[88] The values of the partial molar volume for the sodium alkyl sulfate group surfactant change from positive to negative with increasing pressure, suggesting the compressibility of the micellar surfactants is greater than that of the intermicellar monomer.[89] This also indicates that formation of a hydrophobic bond accompanies a positive volume change at lower pressure, whereas at higher pressure it has a negative value. Such a change is not observed for nonionic surfactants of the polyoxethylene alkyl ether type. This is due to the fact that the partial molar volume of the ethylene oxide group in the micellar state increases slightly with pressure, thereby resisting the compression.[90] Thus, the CMC value increases monotonically and then levels off at high pressure. At the same time, the micellar aggregation number decreases monotonically with pressure for nonionic surfactants,[86,87] although the number for ionic surfactants passes through a minimum at around 1,000 atm.[90]

The micelle size distribution shows a remarkable trend with variation of the surfactant concentration in the solution. The micellar aggregation number is also significantly affected by temperature and the presence of additives in the solution.[91–93] The micelle responds to the changes on addition of additives by growing until the surface area per head group is the same as that of the pure micelle.

3.7 Counterion Binding to Micelles

Counterions present in the solution are expected to be distributed around the charged micelle in two regions: the Stern layer and the diffused double layer. The distribution of counterions in the diffused double layer can be described by the Poisson-Boltzmann equation. The net charge in the diffused double layer must be equal and of opposite sign to the net charge of the micelle (including the Stern layer) to hold electroneutrality of the solution. There are various models present in literature that account for the prediction of the relative counterion concentration in the single or mixed surfactant system as a function of distance between the Stern layer and diffused double layer.[94–97]

Micelles of ionic surfactants are aggregates composed of a compressive core surrounded by a less compressive surface structure[98] and with a rather fluid environment.[99] Copper ions attached to micelles have essentially the same hydration shell near the micellar surface as in the bulk phase, and do not penetrate into the nonpolar part of the micelle.[100] In addition, it is known that the volume change caused by binding of divalent metal ions to micelles is very small.[101] The rate of rotation of the hydrated Na^+ at the micellar surface is unlikely to change by more than 35% upon adsorption from the bulk to the Stern layer of SDS micelles.[102] The gross features of the counterion binding or distribution between the micelle and the bulk solution can be understood in simplified electrostatic models.[103–107] Considering a theory suggested by Evans and Ninham[105] based on the Poisson-Boltzmann equation of potential charge distribution around a spherical micelle, the following result is obtained:

$$\frac{4\pi e^2 (1-\beta)}{\varepsilon \kappa a k T} = 2\sinh(y_0/2)\left\{1+\frac{4}{\kappa R}\left[\frac{\cosh(y_0/2)-1}{\sinh^2(y_0/2)}\right]\right\}^{\frac{1}{2}} \qquad (3.11)$$

where β is the degree of micellar counterion binding quantified by the fraction of free ions in the bulk phase (diffuse layer) and the bound ions in the micelle phase (Stern layer), a is the area per surfactant molecule at the micellar surface, and y_0 is the surface potential, which is a function of κ, a, and β.

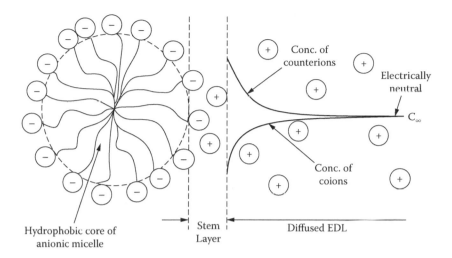

FIGURE 3.7

A schematic of the electric double layer in the vicinity of an adsorption monolayer of ionic surfactant. The stern layer consists of adsorbed immobilized counterions, whereas the diffused layer contains free ions involved in Brownian motion. Near the charged surface there is an accumulation of counterions and a depletion of co-ions, which are having a bulk concentration C_∞.[108]

The spatial charge distribution in the vicinity of the micelle and in the EDL has been demonstrated in Figure 3.7.

Physicochemical characteristics and dynamics of counterion association and dissociation are obtained from the fluorescent quenching method[109–111] or by measuring the relaxation time of nuclear magnetic resonance spectroscopy.[112] Information on micellar counterion binding can also be obtained by conductivity,[113] electromotive force,[114] light scattering,[115] and small-angle neutron scattering measurements.[116] The dynamic state of the ion in a micellar solution can be monitored by its nuclear magnetic resonance spectroscopy using spin-lattice (T_1) and spin-spin (T_2) relaxation times. It is expected that the relaxation times will decrease when the motion of the ion is restricted by binding to a macromolecular surface, compared with the motion of a rotating free monomer state. The degree of micellar counterion binding (β) has been quantified by a fraction of "free" ions in the bulk phase (diffuse layer) and the "bound" ions in the micelle phase (Stern layer).[117]

During competitive binding between a monovalent and divalent cation, the higher valence binding decreases while the cation having lower valence increases; however, the total counterion binding remains approximately constant,[118] equal to 0.65, as represented in Figure 3.8. Competitive

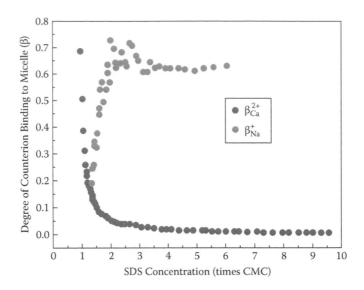

FIGURE 3.8 (See color insert.)
Competitive counterion binding for 0.3 mM of CaCl$_2$ solution at 30°C.

adsorption to the micelle surface has been already investigated in detail.[119–121]

The surfactant micelles are considered to have a highly nonpolar interior and a relatively polar interfacial region. The interior of the micelle is generally the locus of solubilization for very nonpolar solubilizates such as n-alkanes. Solubilizate molecules of relatively high polarity such as alcohols are believed to be solubilized in the interfacial region of the micelle so that their polar functional groups (OH$^-$, for example) can retain their contact with water. However, for molecules such as aromatic hydrocarbons, which are slightly polar, different conflicting propositions have been presented in the literature[121,123] concerning their location in the micelle. Eriksson and Gillberg[124] carried out ^{1}H NMR chemical shift measurements in cetyltrimethylammonium bromide (CTAB) micellar solutions containing benzene as solubilizate. It is observed that small amounts of benzene had a clear effect on the shift of protons near the polar group of the surfactant. This is interpreted to imply that, at low molar solubilization ratios, benzene is located close to the micelle-water interface. Fendler and Patterson[125] investigated the rates of hydrated electron attachment to benzene in micellar solutions using pulse radiolysis. They have inferred that benzene is located in the interfacial region in CTAB micelles, whereas in SDS micelles, it is located in the interior. Investigations on UV spectra concluded that the environments of benzene in both CTAB and SDS micelles are similar, and also that in both the systems benzene is predominantly located in the interior of the micelles.[126] Hirose and Sepulveda[127] have measured the transfer free energies of benzene from water to CTAB and SDS micelles and found them to be equal to the transfer free energy of benzene from water to heptane. On this basis, they inferred that benzene must be located in the interior rather than in the interfacial region of both SDS and CTAB micelles. In general, the fractional distribution of benzene between the interfacial and the interior sites is influenced by the polar head group of the surfactant as well as the extent of benzene molecules solubilized in the micelle.[128]

3.8 Effects on Micelle Formation

Numerous data on counterion binding lead to the following conclusions. The binding of alkali ions to anionic surfactants increases in the order $Li^+ < Na^+ < K^+ < Rb^+ < CS^+$, and binding of anionic counterions to cationic micelles increases in the order $F^- < Cl^- < Br^- < NO_3^- < I^-$. In addition, the counterion binding increases with increasing counterion hydrophobicity,[129–132] promoting micelle formation. For organic counterions, in particular, the binding of monovalent counterions with more than three methylene groups[130–132] and of divalent counterions with more than six methylene groups increases steeply.[133,134] Increase in the length of the alkyl chain initially hinders micelle formation, but longer chains are markedly significant in lowering the CMC and in increasing the aggregation number, responsible for enhanced hydrophobic interaction between the counterion and the micellar core. In general, counterion binding increases with increasing alkyl chain length for ionic surfactants[135–137] and is decreased by addition of alcohols.[138–140] Divalent counterions, on the other hand, stimulate the growth of micelles, and this increase in growth is accompanied by both a decrease in the CMC of almost one order of magnitude compared with monovalent counterions for identical surfactant ions[141–146] and an increase of counterion binding up to more than 0.9.[147] The effects of cationic surfactant and cationic salts counterion binding to SDS, on its micelle size, are presented in Table 3.1.

The stability of the micelles in the presence of counterions is also an important factor. The standard Gibbs free energy gives insight into the stability of the micelles, given the fact that the CMC of the solution is greatly altered by the presence of counterions and the microenvironment of the solution. Detailed theoretical analysis of the micelle stability in the presence of counterions, considering the equilibrium conformational changes from rod to spherical shape and vice versa, has been carried out by Ikeda.[148] The counterions have a significant effect on the size of the anionic,[149] cationic,[150] and nonionic[151] surfactant micelles. Table 3.2 provides the values of the standard free energies with NaCl as the counterion for both cationic and anionic micelles.[152]

TABLE 3.1

Effects of (a) Counterion (Cations) and (b) Cationic Surfactant Binding on SDS Micelle Size at Temperature 20°C[90]

(a) Effect of Cationic Counterions on the Micellar Aggregation Number and Size

SDS Concentration (mM)	Cation Concentration (mM)	Micellar Aggregation Number (SDS)	Diameter of the Micelle (nm)
50.0	0	67.5	3.8
$CaCl_2$			
49.9	1.0	74.1	3.7
49.8	2.0	77.0	3.74
49.7	3.0	80.5	3.8
49.5	5.0	92.1	3.96
$MgCl_2$			
49.8	4.0	79.3	3.78
49.6	7.9	87.0	3.9
49.3	13.8	97.5	4.04
49.0	19.6	95.5	4.02
$Pb(NO_3)_2$			
59.4	1.0	69.5	3.62
58.8	2.0	73.1	3.68
$CdCl_2$			
49.9	1.0	68.7	3.6
49.5	5.0	76.7	3.74
49.0	9.8	80.3	3.8
48.5	14.6	83.7	3.84
47.2	28.3	86.5	3.9
45.5	45.5	81.8	3.82
41.7	83.3	76.1	3.72
$Zn(ClO_4)_2$			
57.7	3.84	71.0	3.64
55.5	7.40	79.3	3.78
53.6	10.7	81.4	3.82
$TbCl_3$			
48.0	2.0	68.8	3.6
46.0	3.7	71.7	3.66
44.6	5.3	71.3	3.66
43.8	6.0	73.7	3.68
41.6	8.3	76.3	3.74
39.6	10.0	81.0	3.88

TABLE 3.1 (CONTINUED)

Effects of (a) Counterion (Cations) and (b) Cationic Surfactant Binding on SDS Micelle Size at Temperature 20°C[90]

(a) Effect of Cationic Counterions on the Micellar Aggregation Number and Size

SDS Concentration (mM)	Cation Concentration (mM)	Micellar Aggregation Number (SDS)	Diameter of the Micelle (nm)
$Y(NO_3)_3$			
48.0	2.0	73.1	3.68
45.4	4.5	80.8	3.8
43.1	6.9	81.0	3.8

(b) Effect of Organic Surfactant as Counterions, on the Micellar Aggregation Number of Both

SDS Concentration (mM)	Organic Surfactant Concentration (mM)	Micellar Aggregation Number (SDS)	Micellar Aggregation Number (counterion)	Diameter of the Micelle (nm)
Octylammonium Chloride				
57.7	3.84	71.1	5.2	3.70
55.5	7.40	80.8	11.8	3.92
53.6	10.70	87.9	19.4	4.10
Decylammonium Chloride				
58.8	1.0	69.1	1.3	3.64
57.7	2.0	72.7	2.8	3.72
55.5	3.7	76.8	5.6	3.82
54.5	4.5	80.2	7.3	3.90
Dodecyltrimethylammonium Chloride				
57.7	0.96	68.5	1.2	3.62
55.6	1.85	70.8	2.6	3.68
53.6	2.67	73.8	4.1	3.76
50.0	4.16	77.4	7.1	3.86
Tetradecyltrimethylammonium Chloride				
58.8	1.0	74.2	1.4	3.62
57.7	2.0	80.2	3.0	3.86
55.5	3.7	82.2	6.0	3.94
Hexadecyltrimethlyammonium Chloride				
48.0	0.38	75.3	0.67	3.72
46.3	0.74	73.5	1.32	3.72
43.1	1.38	77.7	2.82	3.82

TABLE 3.2

Standard Gibbs Free Energy of the Micelle Formation
of (a) SDS and (b) CPC at 303 K

(a) SDS

NaCl (M)	CMC (mM)	$-\Delta G_m^0/KJ(mol^{-1})$
0	8.2	22.4
0.02	3.75	24.2
0.05	2.3	25.4
0.1	1.75	26.2
0.2	0.9	27.8
0.4	0.56	29.1
0.8	0.31	30.6
(b) CPC		
0	0.96	27.3
0.005	0.4	29.8
0.01	0.2	31.3
0.02	0.16	31.9
0.04	0.11	32.9
0.1	0.07	33.9

References

1. Tadros, F. 1984. *The surfactants*. London: Academic Press.
2. Lange, K.R. 1999. *Surfactants: A practical handbook.* Cincinnati: Hanser Gardner Publications.
3. Tadros, T.F. 2005. *Applied surfactants: Principles and applications.* Weinheim: Wiley VCH GmbH & Co. KGaA.
4. Robb, I.D. 1997. *Specialist surfactants.* London: Blackie Academic and Professional, Chapman & Hall.
5. Myers, D. 2006. *Surfactant science and technology.* Hoboken, NJ: Wiley Interscience.
6. McBain, J.W. 1950. *Colloid science.* Boston: Heath.
7. Harkins, W.D., Mattoon, R.W., and Corrin, M.L. 1946. Structure of soap micelles as indicated by x-rays and interpreted by the theory of molecular orientation. *J. Colloid Sci.* 1: 105–126.
8. Berberich, K.A., Reinsborough, V.C., and Shaw, C.N. 2000. Kinetic and solubility studies in zwitterionic surfactant solutions. *J. Solution Chem.* 29: 1017–1026.
9. Lopez-Diaz, D., and Castillo, R. 2010. The wormlike micellar solution made of a zwitterionic surfactant (TDPS), an anionic surfactant (SDS), and brine in the semidilute regime. *J. Phys. Chem. B* 114: 8917–8925.
10. McLachlan, A.A., and Marangoni, D.G. 2006. Interactions between zwitterionic and conventional anionic and cationic surfactants. *J. Colloid Interf. Sci.* 295: 243–248.

11. Rosen, M.J. 2004. *Surfactants and Interfacial phenomena*. Hoboken, NJ: John Wiley & Sons.
12. Debye, P., and Anaker, E.W. 1951. Micelle shape from dissymmetry measurements. *J. Phys. Colloid Chem.* 55: 644–655.
13. Balzer, D., and Luders, H. 2000. *Nonionic surfactants: Alkyl polyglucosides*. New York: Marcel Dekker.
14. Anainsson, E.A.G., and Wall, S.N. 1975. Kinetics of step-wise micelle association. Correction and improvement. *J. Phys. Chem.* 79: 857–858.
15. Barry, B.W., and Eini, D.I.D. 1976. Surface properties and micelle formation of long-chain polyoxyethylene nonionic surfactants. *J. Colloid Interf. Sci.* 54: 339–347.
16. Rosen, M.J. 1975. Relationship of structure to properties in surfactants. III. Adsorption at the solid-liquid interface from aqueous solution. *J. Am. Oil Chem. Soc.* 52: 431–435.
17. Esumi, K., and Uneno, M. 1997. *Structure-performance relationship in surfactants*. Surfactant Science Series V. Boca Raton, FL: CRC Press.
18. Mittal, K.L., and Shah, D.O. 2003. *Adsorption and aggregation of surfactants in solution*. New York: Marcel Dekker.
19. Schramm, L.L., and Smith, R.G. 1985. The influence of natural surfactants on interfacial charges in the hot-water process for recovering bitumen from the athabasca oil sands. *Colloids Surf.* 14: 67–85.
20. Takamura, K. 1982. Microscopic structure of athabasca oil sand. *Can. J. Chem. Eng.* 60: 538–545.
21. Isaacs, E.I., and Chow R.S. 1992. *Emulsions: Fundamentals and applications in the petroleum industry*, ed. L.L. Schramm. Washington, DC: American Chemical Society.
22. Tanford, C. 1980. *The Hydrophobic effect: Formation of micelles and biological membranes*. New York: John Wiley & Sons.
23. Lindman, B. 1984. Structural aspects of surfactant micellar systems. In *Surfactants*, ed. Th.F. Tadros, 83–111. London: Academic Press.
24. Tanford, C. 1972. Micelle shape and size. *J. Phys. Chem.* 76: 3020–3024.
25. Tanford, C. 1974. Theory of micelle formation in aqueous solution. *J. Phys. Chem.* 78: 2469–2479.
26. Tanford, C. 1974. Thermodynamics of micelle formation: Prediction of micelle size and size distribution. *Proc. Natl. Acad. Sci. USA* 71: 1811–1815.
27. Frank, H.S., and Evans, M.W. 1945. Free volume and entropy in condensed systems III. Entropy in binary liquid mixtures; partial molal entropy in dilute solutions; structure and thermodynamics in aqueous electrolytes. *J. Chem. Phys.* 13:507–532.
28. Clint, J.H. 1992. *Surfactant aggregation*. New York: Chapman & Hall.
29. Miyajima, K., Lee, T., and Nakagaki, M. 1984. Micelle formation of polyoxyethylene cholesteryl ether in water. *Chem. Pharma. Bull.* 32: 3670–3673.
30. Greiss, W. 1955. About the relationship between the constitution and properties of alkyl benzene, each with a straight or branched alkyl chain up to 18 carbon atoms I. *Fette Seifen Anstrichmi.* 57: 24–32.
31. Galembeck, F. 2004. *Surface and colloid science*. Berlin: Springer-Verlag.
32. Evans, H.C. 1956. Alkyl sulfates. Part I. Critical micelle concentrations of the sodium salts. *J. Chem. Soc.* 80: 579–586.
33. Stigter, D. 1974. Micelle formation by ionic surfactants. II. Specificity of head groups, micelle structure. *J. Phys. Chem.* 78: 2480–2485.

34. Tokiwa, F., and Tsuiji, K. 1972. NMR study of the interaction between anionic and nonionic surfactants in their mixed micelles. II. *J. Colloid Interf. Sci.* 41: 343–349.
35. Schick, M.J. 1962. Surface films of nonionic detergents. I. Surface tension study. *J. Colloid Sci.* 17: 801–813.
36. Haverd, V.E., and Warr, G.G. 2000. Cation selectivity at air/anionic surfactant solution interfaces. *Langmuir* 16: 157–160.
37. Flockhart, B.D. 1961. The effect of temperature on the critical micelle concentration of some paraffin-chain salts. *J. Colloid Sci.* 16: 484–492.
38. Mujamoto, S. 1960. The effect of metallic ions on surface chemical phenomena. IV. Surface tension measurement on aqueous solutions of metal dodecyl sulfates. *Bull. Chem. Soc. Jpn.* 33: 375–379.
39. Tori, K., and Nakagawa, T. 1963. Colloid chemical properties of ampholytic surfactants V. Temperature and salt effects on the critical micelle concentration of long-chain alkyl betaine. *Colloid Polym. Sci.* 189: 50–55.
40. Dahanayake, M., and Rosen, M.J. 1984. *Structure/performance relationships in surfactants*, ed. M.J. Rosen. ACS Symposium Series 253. Washington, DC: American Chemical Society.
41. Jones, E.R., and Bury, C.R. 1927. The freezing-points of concentrated solutions. Part II. Solutions of formic, acetic, propionic, and butyric acids. *Philos. Mag.* 4: 841–848.
42. Shinoda, K., and Hutchinson, E. 1962. Pseudo-phase separation model for thermodynamic calculations on micellar solutions. *J. Phys. Chem.* 66: 577–582.
43. Fisher, L.R., and Oakenfull, D.G. 1977. Micelles in aqueous solution. *Chem. Soc. Rev.* 6: 25–42.
44. Mukerjee, P. 1972. Size distribution of small and large micelles. Multiple equilibrium analysis. *J. Phys. Chem.* 76: 565–570.
45. Emerson, M.F., and Holtzer, A. 1965. On the ionic strength dependence of micelle number. *J. Phys. Chem.* 69: 3718–3721.
46. Shinoda, K., Nakagawa, T., Tamamushi, B., et al. 1962. *Colloidal surfactants*. New York: Academic Press.
47. Moroi, Y. 1992. *Micelles, theoretical and applied aspects*. New York: Plenum Press.
48. Maeda, H. 1995. A simple thermodynamic analysis of the stability of ionic/nonionic mixed micelles. *J. Colloid. Interf. Sci.* 172: 98–105.
49. Goldsipe, A., and Blankschtein, D. 2007. Molecular-thermodynamic theory of micellization of multicomponent surfactant mixtures. 1. Conventional (pH-insensitive) surfactants. *Langmuir* 23: 5942–5952.
    Goldsipe, A., and Blankschtein, D. 2007. Molecular-thermodynamic theory of micellization of multicomponent surfactant mixtures. 2. pH-sensitive surfactants. *Langmuir* 23: 5953–5962.
50. Stephenson, B.C., Stafford, K.A., Beers, K.J., and Blankschtein, D. 2008. Application of computer simulation free-energy methods to compute the free energy of micellization as a function of micelle composition. 1. Theory. *J. Phys. Chem. B* 112: 1634–1640.
    Stephenson, B.C., Stafford, K.A., Beers, K.J., and Blankschtein, D. 2008. Application of computer simulation free-energy methods to compute the free energy of micellization as a function of micelle composition. 2. Implementation. *J. Phys. Chem. B* 112: 1641–1656.

51. Stephenson, B.C., Goldsipe, A., Beers, K.J., and Blankschtein, D. 2007. Quantifying the hydrophobic effect. 2. A computer simulation–molecular-thermodynamic model for the micellization of nonionic surfactants in aqueous solution. *J. Phys. Chem. B* 111: 1045–1062.
52. Hartley, G.S. 1926. *Aqueous solutions of paraffin-chain salts.* Paris: Hermann.
53. Reich, I. 1956. Factors responsible for the stability of detergent micelles. *J. Phys. Chem.* 60: 257–262.
54. McBain, J.W. 1944. In *Colloid chemistry, theoretical and applied,* ed. J. Alexander. New York: Reinhold. 102.
55. Mattoon, R.H., Stearns, R.S., and Harkins, W.D. 1947. Structure for soap micelles as indicated by a previously unrecognized x-ray diffraction band. *J. Chem. Phys.* 15: 209–210.
    Harkins, W.D. 1948. A cylindrical model for the small soap micelle. *J. Chem. Phys.* 16: 156–157.
56. Debye, P., and Anacker, E.W. 1951. Micelle shape from dissymmetry measurements. *J. Phys. Chem.* 55: 644–655.
57. Hayter, J.B., and Penfold, J. 1981. Self-consistent structural and dynamic study of concentrated micelle solutions. *J. Chem. Soc. Faraday Trans. 1* 77: 1851–1863.
58. Gruen, D.W.R. 1985. A model for the chains in amphiphilic aggregates. 1. Comparison with a molecular dynamics simulation of a bilayer. *J. Phys. Chem.* 89: 146–153.
    Gruen, D.W.R. 1985. A model for the chains in amphiphilic aggregates. 2. Thermodynamic and experimental comparisons for aggregates of different shape and size. *J. Phys. Chem.* 89: 153–163.
59. Cabane, B., Duplessix, R., and Zemb, T. 1985. High resolution neutron scattering on ionic surfactant micelles: SDS in water. *J. Phys. (Paris)* 46: 2161–2171.
60. Courchene, W.L. 1964. Micellar properties from hydrodynamic data. *J. Phys. Chem.* 68: 1870–1874.
61. Kang, K.-H., and Lim, K.-H. 2011. A SANS study for structural transition of micelles of cationic octadecyl trimethyl ammonium chloride and anionic ammonium dodecyl sulfate surfactants in aqueous solutions. *Colloids. Surf. A Physicochem. Eng. Asp.,* doi:10.1016/j.colsurfa.2011.05.060.
62. Bendedouch, D., Chen, S.-H., and Koeler, W.C. 1983. Determination of interparticle structure factors in ionic micellar solutions by small angle neutron scattering. *J. Phys. Chem.* 87: 2621–2628.
63. Chevalier, Y., and Chachaty, C. 1984. NMR investigation of the micellar properties of monoalkylphoshpates. *Colloid Polym. Sci.* 262: 489–496.
64. Kawaguchi, T., Hamanaka, T., and Mitsui, T. 1983. X-ray structural studies of some nonionic detergent micelles. *J. Colloid Interf. Sci.* 96: 437–453.
65. Birdi, K.S. The size, the shape and hydration of micelles in aqueous medium. *Prog. Colloid Polym. Sci.* 70: 20–23.
66. Kauzmann, W. 1959. Some factors in the interpretation of protein denaturation. *Adv. Protein Chem.* 44: 1–66.
67. Ben-Nairn, A. 1980. *Hydrophobic interactions.* New York: Plenum Press.
68. Evans, D.F., Allen, M., Ninham, B.W., and Fouda, A. 1984. Critical micelle concentrations for alkyltrimethylammonium bromides in water from 25 to 160°C. *J. Solution Chem.* 13: 87–101.
69. Evans, D.F., and Wightman, P.J. 1982. Micelle formation above 100°C. *J. Colloid Interf. Sci.* 86: 515–524.

70. Archer, D.G., Albert, H.J., White, D.E., and Wood, R.H. 1984. Enthalpies of dilution and heat capacities of aqueous solutions of sodium n-dodecyl sulfonate and sodium 4-(1-metthlundecyl) benzene sulfate from 347 to 451 K. *J. Colloid Interf. Sci.* 100: 68–81.
71. Archer, D.G. 1987. Enthalpies of dilution of aqueous tetradecyltrimethylammonium bromide from 50 to 175°C. *J. Solution Chem.* 16: 347–365.
72. Jones, M.N., and Piercy, J. 1972. Light scattering studies on n-dodecyltrimethylammonium bromide and n-dodecylpyridinium iodide. *J. Chem. Soc. Faraday Trans. 1* 68: 1839–1849.
73. Croonen, Y., Gelade, E., Van der Zegel, M., et al. 1983. Influence of salt, detergent concentration, and temperature on the fluorescence quenching of 1-methylpyrene in sodium dodecyl sulfate with m-dicyanobenzene. *J. Phys. Chem.* 87: 1426–1431.
74. Benjamin, L. 1964. Calorimetric studies of the micellization of dimethyl-n-alkylamine oxides. *J. Phys. Chem.* 68: 3575–3581.
75. Malliaris, A., Moigne, J.L., Sturm, J., and Zana, R. 1985. Temperature dependence of the micelle aggregation number and rate of intramicellar excimer formation in aqueous surfactant solutions. *J. Phys. Chem.* 89: 2709–2713.
76. Schick, M.J. 1967. *Nonionic surfactants.* New York: Marcel Dekker.
77. Brown, W., Johnson, R., Stilbs, P., and Lindman, B. 1983. Size and shape of nonionic amphiphile (C12E6) micelles in dilute aqueous solutions as derived from quasielastic and intensity light scattering, sedimentation, and pulsed-field-gradient nuclear magnetic resonance self-diffusion data. *J. Phys. Chem.* 87: 4548–4553.
78. Zulauf, M., and Rosenbusch, J.P. 1983. Micelle clusters of octylhydroxyoligo(oxyethylenes). *J. Phys. Chem.* 87: 856–862.
79. Cebula, D.J., and Ottewill, R.H. 1982. Neutron scattering studies on micelles of dodecylhexaoxyethylene glycol monoether. *Colloid Polym. Sci.* 260: 1118–1120.
80. Kresheck, G.C. 1975. In *Water: A comprehensive treatise: Aqueous solutions of amphiphiles and macromolecules,* ed. F. Franks, 96–98. Vol. 4. New York: Plenum Press.
81. Yamanaka, M., Aratono, M., Motomura, K., and Matuura, R. 1984. Effect of pressure on the adsorption and micelle formation of aqueous dodecyltrimethylammonium chloride solution-hexane system. *Colloid Polym. Sci.* 262: 338–341.
82. Ljosland, E., Blokhus, A.M., Veggeland, K., Backlund, S., and Hoiland, H. 1985. Shape transitions in aqueous micellar systems as a function of pressure and temperature. *Prog. Colloid Polym. Sci.* 70: 34–37.
83. Yamanaka, M., Aratono, M., and Motomura, K. 1986. Effect of pressure on the adsorption and micelle formation of aqueous dodecyltrimethylammonium chloride–cyclohexane system. *Bull. Chem. Soc. Jpn.* 59: 2695–2698.
84. Ikawa, Y., Tsuru, S., Murata, Y., Okawauchi, M., Shigematsu, M., and Sugihara, G. 1988. A pressure and temperature study on solubility and micelle formation of sodium perfluorodecanoate in aqueous solution. *J. Solution Chem.* 17: 125–137.
85. Nishikido, N., Yoshimura, N., Tanaka, M., and Kaneshina, S. 1980. Effect of pressure on the solution behavior of nonionic surfactants in water. *J. Colloid Interf. Sci.* 78: 338–346.
86. Nishikido, N., Shinozaki, M., Sugihara, G., and Tanaka, M. 1981. A study on the micelle formation of surfactants in aqueous solutions under high pressure by laser light-scattering technique II. *J. Colloid Interf. Sci.* 82: 352–361.

87. Okawauchi, M., Shinoazki, M., Ikawa, Y., and Tanaka, M. 1987. A light scattering study on micelle formation of nonionic surfactants as a function of concentration and pressure by applying small system thermodynamics. *J. Phys. Chem.* 91: 109–112.

88. Kaneshina, S., Tanaka, M., Tomida, T., and Matuura, R. 1974. Micelle formation of sodium alkylsulfate under high pressures. *J. Colloid Interf. Sci.* 48: 450–460.

89. Tanaka, M., Kaneshina, S., Shin-no, K., Okajima, T., and Tomida, T. 1974. Partial molal volumes of surfactant and its homologous salts under high pressure. *J. Colloid Interf. Sci.* 46: 132–138.

90. Nishikido, N., Shinozaki, M., Sugihara, G., Tanaka, M., and Kaneshina, S. 1980. A study on the micelle formation of surfactants in aqueous solutions under high pressure by laser light-scattering technique I. *J. Colloid Interf. Sci.* 74: 474–482.

91. Almgren, M., and Swarup, S. 1983. Size of sodium dodecyl sulfate micelles in the presence of additives I. Alcohols and other polar compounds. *J. Colloid Interf. Sci.* 91: 256–266.

92. Almgren, M., and Swarup, S. 1982. Size of sodium dodecyl sulfate micelles in the presence of additives. 2. Aromatic and saturated hydrocarbons. *J. Phys. Chem.* 86: 4212–4216.

93. Almgren, M., and Swarup, S. 1983. Size of sodium dodecyl sulfate micelles in the presence of additives. 3. Multivalent and hydrophobic counterions, cationic and nonionic surfactants. *J. Phys. Chem.* 87: 876–881.

94. Stigter, D. 1964. On the adsorption of counterions at the surface of detergent micelles. *J. Phys. Chem.* 68: 3603–3611.

95. Stigter, D. 1975. Micelle formation by ionic surfactants. III. Model of Stern layer, ion distribution, and potential fluctuations. *J. Phys. Chem.* 79: 1008–1014.

96. Beunen, J.A., and Ruckenstein, E. 1983. A model for counterion binding to ionic micellar aggregates. *J. Colloid Interf. Sci.* 96: 469–487.

97. Rathman, J.F., and Scamehorn, J.F. 1984. Counterion binding on mixed micelles. *J. Phys. Chem.* 88: 5807–5816.

98. Bloor, D.M., Gormally, J., and Wyn-Jones, E.J. 1984. Adiabatic compressibility of surfactant micelles in aqueous solutions. *Chem. Soc. Faraday Trans. 1* 80: 1915–1923.

99. Stark, R.E., Kasakevich, M.L., and Granger, J.W. 1982. Molecular motion of micellar solutes. A carbon-13 NMR relaxation study. *J. Phys. Chem.* 86: 335–340.

100. Narayana, P.A., Li, W. A.S., and Kevan, L. 1982. A new method for detection of preferential adsorption of metal cations at anionic micellar surfaces in frozen aqueous solutions by electron spin echo spectrometry. *J. Phys. Chem.* 86: 3–5.

101. Baumuller, W., Hoffman, H., Ulbricht, W., Tondre, C., and Zana, R. 1978. Chemical relaxation and equilibrium studies of aqueous solutions of laurylsulfate micelles in the presence of divalent metal ions. *J. Colloid Interf. Sci.* 64: 418–437.

102. Robb, I.D. 1971. The binding of counter ions to detergent micelles: The nature of the Stern layer. *J. Colloid Interf. Sci.* 37: 521–527.

103. Gunnarsson, G., Jonsson, B., and Wennerstrom, H. 1980. Surfactant association into micelles. An electrostatic approach. *J. Phys. Chem.* 84: 3114–3121.

104. Linse, P., Gunnarsson, G., and Jonsson, B. 1982. Electrostatic interactions in micellar solutions. A comparison between Monte Carlo simulations and solutions of the Poisson-Boltzmann equation. *J. Phys. Chem.* 86: 413–421.

105. Evans, D.F., and Ninham, B.W. 1983. Ion binding and the hydrophobic effect. *J. Phys. Chem.* 87: 5025–5032.
106. Shirahama, K. 1974. Ionic equilibria in micellar solutions. *Bull. Chem. Soc. Jpn.* 47: 3165–3166.
107. Kaneko, T. 1986. A reinterpretation for ion association at micellar surfaces due to an electrostatic model. *Bull. Chem. Soc. Jpn.* 59: 1290–1292.
108. Kralchevsky, P.A., Danov, K.D., Broze, G., and Mehreteab, A. 1999. Thermodynamics of ionic surfactant adsorption with account for the counterion binding: Effect of salts of various valency. *Langmuir* 15: 2351–2365.
109. Singer, L.A. 1982. *Fluorescent probes of micellar systems—An overview, solution behaviour of surfactants*, 73–112. New York: Plenum Press.
110. Grieser, F., and Drummond, C.J. 1988. The physicochemical properties of self assembled surfactant aggregates as determined by some molecular spectroscopic probe techniques. *J. Phys. Chem.* 95: 5580–5593.
111. Asakawa, T., Kubode, H., Ozawa, T., Ohta, A., and Miyagishi, S. 2005. Micellar counterion binding probed by fluorescence quenching of 6-methoxy-N-(3-sulfopropyl) quinolinium. *J. Oleo Sci.* 54: 545–552.
112. Yoshida, T., Taga, K., Okabayashi, H., Matsushita, K., Kamaya, H., and Ueda, I. 1986. Counterion binding to micelles measured by sodium-23 NMR relaxation times: Enhancement by anesthetics. *J. Colloid Interf. Sci.* 109: 336–340.
113. Shanks, P.C., and Franses, E.J. 1992. Estimation of Micellization parameters of aqueous sodium dodecyl sulfate from conductivity data. *J. Phys. Chem.* 96: 1794–1799.
114. Gallon, L., Lelievre, J., and Gaboriaud, R. 1999. Counterion effects in aqueous solutions of cationic surfactants: Electromotive force measurements and thermodynamic model. *J. Colloid Interf. Sci.* 213: 287–297.
115. Fujio, K., and Ikeda, S. 1991. Size and spherical micelles of dodecylpyridinium bromide in aqueous NaBr solutions. *Langmuir* 7: 2899–2904.
116. Quirion, F., and Magid, L.J. 1986. Growth and counterion binding of cetltrimethylammonium bromide aggregates at 25°C: A neutron and light scattering study. *J. Phys. Chem.* 90: 5435–5441.
117. Asakawa, T., Kubode, H., Ozawa, T., Ohta, A., and Miyagishi, S. 2005. Micellar counterion binding probed by fluorescence quenching of 6-methoxy-N-(3-sulfopropyl) quinolinium. *J. Oleo Sci.* 54: 545–552.
118. Maneedaeng, A., Haller, K.J., Brian, B.P., Grady, P., and Flood, A.E. 2011. Thermodynamic parameters and counterion binding to the micelle in binary anionic surfactant systems. *J. Colloid Interf. Sci.* 356: 598–604.
119. Koshinuma, M. 1983. Ionic composition of micelles and the dissolved state of the mixed surfactant solutions of calcium and sodium dodecyl sulfates. *Bull. Chem. Soc. Jpn.* 56: 2341–2347.
120. Paton-Morales, P., and Talens-Alesson, F.I. 2002. Effect of competitive adsorption of Zn^{2+} on the flocculation of lauryl sulfate micelles by Al^{3+}. *Langmuir* 18: 8295–8301.
121. Samper, E., Rodríguez, M., De la Rubia, M.A., and Pratsa, D. 2009. Removal of metal ions at low concentration by micellar-enhanced ultrafiltration (MEUF) using sodium dodecyl sulfate (SDS) and linear alkylbenzene sulfonate (LAS). *Sep. Purif. Technol.* 65: 337–342.
122. Fendler, J.H., and Fendler, E.J. 1975. *Catalysis in micellar and macromolecular systems*. New York: Academic Press.

123. Simon, S.A., McDaniel, R.V., and McIntosh, T.J. 1982. Interaction of benzene with micelles and bilayers. *J. Phys. Chem.* 86: 1449–1456.
124. Eriksson, J.C., and Gillberg, G. 1966. NMR-Studies of the solubilisation of aromatic compounds in cetyltrimethylammonium bromide solution II. *Acta Chemica Scand.* 20: 2019–2027.
125. Patterson, L.K., and Fendler, J.H. 1970. Micellar effects on the reactivity of the hydrated electron with benzene. *J. Phys. Chem.* 74: 4608–4609.
126. Rehfeld, S.J. 1970. Solubilization of benzene in aqueous sodium dodecyl sulfate solutions measured by differential spectroscopy. *J. Phys. Chem.* 74: 117–122.
127. Hirose, C., and Sepulveda, L. 1981. Transfer free energies of p-alkyl-substituted benzene derivatives, benzene, and toluene from water to cationic and anionic micelles and to n-heptane. *J. Phys. Chem.* 85: 3689–3694.
128. Nagarajan, R., Chaiko, M.A., and Ruckenstein, E. 1984. Locus of solubilization of benzene in surfactant micelles. *J. Phys. Chem.* 88: 2916–2922.
129. Mukerjee, P., Mysels, K.J., and Kapauan, P. 1967. Counterion specificity in the formation of ionic micelles—Size, hydration, and hydrophobic bonding effects. *J. Phys. Chem.* 71: 4166–4175.
130. Anacker, E.W., and Underwood, A.L. 1981. Organic counterions and micellar parameters: *n*-Alkyl carboxylates. *J. Phys. Chem.* 85: 2463–2466.
131. Underwood, A.L., and Anacker, E.W. 1984. Organic counterions and micellar parameters: Methyl-, chloro-, and phenyl-substituted acetates. *J. Colloid Interf. Sci.* 100: 128–135.
132. Underwood, A.L., and Anacker, E.W. 1984. Organic counterions and micellar parameters: Substituent effects in a series of benzoates. *J. Phys. Chem.* 88: 2390–2393.
133. Moroi, Y., Matuura, R., Kuwamura, T., and Inokuma, S. 1986. Anionic surfactants with divalent counterions of separate electric charge: Solubility and micelle formation. *J. Colloid Interf. Sci.* 113: 225–231.
134. Lissi, E., Abuin, E., Cuccovia, I.M., and Chaimovich, H. 1986. Ion exchange between alkyl dicarboxylates and hydrophilic anions at the surface of cetyltrimethylammonium micelles. *J. Colloid Interf. Sci.* 112: 513–520.
135. Moroi, Y., Nishikido, N., Uehara, H., and Matuura, R. 1975. An interrelationship between heat of micelle formation and critical micelle concentration. *J. Colloid Interf. Sci.* 50: 254–264.
136. Zana, R. 1980. Ionization of cationic micelles: Effect of the detergent structure. *J. Colloid Interf. Sci.* 78: 330–337.
137. Leung, R., and Shah, D.O. 1986. Dynamic properties of micellar solutions. I. Effects of short-chain alcohols and polymers on micellar stability. *J. Colloid. Interf. Sci.* 113: 484–499.
138. Shinoda, K. 1953. The effect of chain length, salts and alcohols on the critical micelle concentration. *Bull. Chem. Soc. Jpn.* 26: 101–105.
139. Larsen, J.W., and Tepley, L.B. 1974. Effect of aqueous alcoholic solvents on counterion-binding to CTAB micelles. *J. Colloid Interf. Sci.* 49: 113–118.
140. Koshinuma, M. 1979. Studies of the dissolved state of sodium tetradecyl sulfate in ethanol–water mixtures by measurements of the activity. *Bull. Chem. Soc. Jpn.* 52: 1790–1795.
141. Hato, M., and Shinoda, K. 1973. The solubilities, critical micelle concentrations, and Krafft points of bivalent metal alkyl sulfates. *Bull. Chem. Soc. Jpn.* 46: 3889–3890.

142. Miyamoto, S. 1960. The effect of metallic ions on surface chemical phenomena. III. Solubility of various metal dodecyl sulfates in water. *Bull. Chem. Soc. Jpn.* 33: 371–375.
143. Satake, I., Iwamatsu, I., Hosokawa, S., and Matuura, R. 1963. The surface activities of bivalent metal alkyl sulfates. I. On the micelles of some metal alkyl sulfates. *Bull. Chem. Soc. Jpn.* 36: 204–209.
144. Satake, I., and Matuura, R. 1963. The surface activities of bivalent metal alkyl sulfates. II. The solubilization of some hydrocarbons in aqueous solutions of bivalent metal alkyl sulfates. *Bull. Chem. Soc. Jpn.* 36: 813–817.
145. Moroi, Y., Motomura, K., and Matuura, R. 1974. The critical micelle concentration of sodium dodecyl sulfate-bivalent metal dodecyl sulfate mixtures in aqueous solutions. *J. Colloid Interf. Sci.* 46: 111–117.
146. Romsted, L.S. 1984. *Micellar effects on reaction rates and equilibria.* New York: Plenum Press.
147. Moroi, Y., and Matuura, R. 1985. Size distribution of anionic surfactant micelles. *J. Phys. Chem.* 89: 2923–2928.
148. Ikeda, S. 1991. Stability of spherical and rod-like micelles of ionic surfactants, in relation to their counterion binding and modes of hydration. *Col. Polym. Sci.* 269: 49–61.
149. Hayashi, S., and Ikeda, S. 1980. Micelle size and shape of sodium dodecyl sulfate in concentrated NaCl solutions. *J. Phys. Chem.* 84: 744–751.
150. Anacker, E.W., and Ghose, H.M. 1968. Counterions and micelle size. II. Light scattering by solutions of cetylpyridinium salts. *J. Am. Chem. Soc.* 90: 3161–3166.
151. Streletzky, K., and Phillies, G.D.J. 1996. Temperature dependence of Triton X-100 micelle size and hydration. *Langmuir* 11: 42–47.
152. Chatterjee, A., Moulik, A.P., Sanyal, S.K., et al. 2001. Thermodynamics of micelle formation of ionic surfactants: A critical assessment for sodium dodecyl sulfate, cetyl pyridinium chloride and dioctyl sulfosuccinate (Na salt) by microcalorimetric, conductometric, and tensiometric measurements. *J. Phys. Chem. B* 105: 12823–12831.

4

Selection of Surfactant

Micellar-enhanced ultrafiltration (MEUF) is a process that involves the use of surfactant to form micelles, which are responsible for binding to pollutants, and hence rejected by a suitable membrane. Micelle formation and micelle characteristics are important phenomena in the entire process. The electrostatic binding of a charged contaminant to the micelle depends largely on the size, distribution of micelles in the solution, and other environmental factors (like pH, temperature, presence of co-ions, etc.). Thus, choice of surfactant is a significant consideration for effective removal of the impurities from the waste stream. By now, it is well understood that the degree of hydrophobic interaction at the micelle-water interface plays a vital role in determining whether the solute would be adsorbed on the surface or in the interior of the micelle. In MEUF, metal cations and inorganic pollutants form a bond with the head of the ionic micelle's surface (see Figure 4.1), which is oppositely charged via electrostatic interaction.[1-4] This mechanism is not applicable for MEUF of organic materials. In MEUF of organic contaminants, dissolved organic solutes will be solubilized within the palisade layer or the core of micelles (tail of the micelles) via the Van der Waals force.[2,5] From an engineering aspect, it is worthy to quantify the minimum amount of surfactant required to produce a desired proportion of adsorption of the pollutant, considering the size, molecular structure, and cost of the surfactant. The key question to address in this chapter would be: How does the degree of solubilization of an organic/inorganic solute affect (or relate to) the molecular structure of a surfactant?

For hydrocarbons and long-chain polar compounds that are solubilized in the interior of the micelle or deep in the palisade layer, the amount of material solubilized generally increases with the size of the micelles. Therefore, any factor that causes an increase in either the diameter of the micelle or its aggregation number can be expected to produce an increase in the solubilization capacity for this type of material. Since aggregation number increases with increase in the degree of dissimilarity between solvent and surfactant, an increase in the chain length of the hydrophobic portion of the surfactant generally results in increased solubilization capacity for hydrocarbons in the interior of the micelle in an aqueous environment. Nonionic surfactants are better solubilizing agents than their ionic counterparts in very dilute solutions, because of their lower critical micelle concentrations. In general, the order of solubilizing power for hydrocarbons and polar compounds that are solubilized in the inner core appears to be as follows:

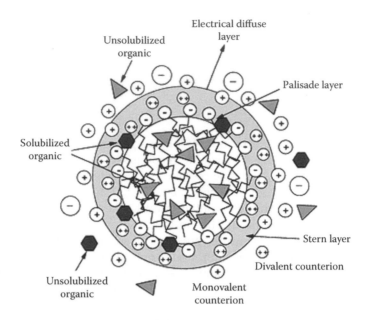

FIGURE 4.1
Solubilization of pollutant molecule in the micelle.

nonionics > cationics > anionics for surfactants with the same hydrophobic chain length.[6-8] The greater solubilizing power of cationics, compared to anionics of equivalent hydrophobic chain length, may be due to looser packing of the surfactant molecules in the micelles of the former.[9,10] Some of the desirable characteristics of an ideal surfactant in MEUF are:

1. High degree of solubilization per unit weight of the surfactant dissolved.
2. Formation of large micelle structure so that the resulting solution can be filtered using highly permeable membranes.
3. Low monomer concentration, so that little surfactant is wasted.
4. Minimal phase separation problems (macroemulsion formation, precipitation, gelling, etc.).
5. The Krafft point temperature for ionic surfactant (the temperature at which the solubility of the surfactant is equal to the critical micellar concentration (CMC)[11]) should be very low. Nonionic surfactants do not have a specific temperature effect on the property curve.
6. The CMC of the surfactant should be as low as possible.

One desirable characteristic of a surfactant is a long hydrocarbon chain, since this results in large micelles, high solubilization, and lower monomer concentration above CMC. Anionic surfactants are restricted to hydrocarbon chain lengths of about 12 carbons or less if the process is to be applied at room temperature. For longer hydrophobic groups, the Krafft temperature is above room temperature; hence the surfactant precipitates and MEUF are ineffective.[11]

Nonionic surfactant-forming micelles can have high solubilization capacities (per mole of surfactant) and have low monomer concentrations in micellar solutions. These surfactants would appear to be good candidates for use in MEUF. The solubilization capacities of the nonionic surfactants are not very high compared to those of anionic or cationic surfactants because of their relatively high molecular weights. In most cases, nonionic surfactants are expensive and are not economically viable for use in MEUF.[12]

Cationic surfactants generally have much lower Krafft temperatures than anionic surfactants of similar hydrophobic group size. Therefore, cationic surfactants with large hydrophobic groups can be used in MEUF, resulting in large micelles, high solubilization capacities, and low monomer concentrations. Cationic surfactants do not pose significant environmental risks.[13] In general, cationic surfactants are the choice of surfactants in MEUF.

Thus, an ideal surfactant would be one that has a small value of CMC, a Krafft point above the room temperature, is structured so that the micelles formed are completely rejected, and does not solubilize any undesired solute. Also, it can be added that considering the limitations from a structural viewpoint, a cosolubilizate can be thought of as increasing the solute rejection capabilities of a UF membrane. Kandori et al. have shown that cosolubilization of phenol in the presence of cyclohexane using CO-850 surfactant is greatly enhanced during MEUF.[14]

4.1 Ionic Surfactant

4.1.1 MEUF Using Cationic Surfactant

The hydrophilic head of cationic surfactants bears a positive charge that makes it possible to attract the anionic contaminants, such as anionic dyes, and other organic contaminants, such as phenols, alcohols, aromatic compounds, etc. Cetylpyridinium chloride (CPC) is one of the potential cationic surfactants for wide use in MEUF. It has been observed that the surfactant with the same hydrophilic group (cetyltrimethylammonium bromide [CTAB] and octyltrimethyl ammonium bromide [OTAB]), the hydrophobic chain length (hydrophobic tail), strongly influences the solubilization of phenol into micelles. The increment of the hydrophobic alkyl chain of the surfactant results in reduction in CMC of surfactant, leading to enhancement of the solubilization. Zaghbani

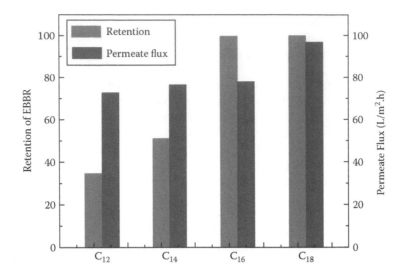

FIGURE 4.2 (See color insert.)
Comparison of permeate flux and retention using different alkyl chain lengths of trimethyl-ammonium bromide for removal of 0.001 (M) EBBR dye at 10 CMC surfactant concentration.

et al.[15] studied the effect of a hydrophobic alkyl chain using n-alkyltrimethylammonium bromide surfactants: dodecyltrimethylammonium bromide (C_{12}TAB), tetradecyltrimethylammonium bromide (C_{14}TAB), cetyltrimethylammonium bromide (C_{16}TAB), and octadecyltrimethylammonium bromide (C_{18}TAB) on Eriochrome Blue Black R (EBBR) dye. From the study, they found that the dye retention increased as the hydrophobic alkyl chain increased (C_{12}TAB (34.6%) < C_{14}TAB (51.1%) < C_{16}TAB (99.6%) < C_{18}TAB (99.9%)) due to hydrophobic interaction, which contributes to the increment in solubilization of dyes by micelles, as presented in Figure 4.2.

4.1.2 Anionic Surfactant

The micelle hydrophobe size has a direct correlation with the retention of contaminants as well as permeate flux. The surfactant separation efficiency is dependent on its value of CMC. However, the separation is also enhanced due to the effects of concentration polarization. In concentration polarization, the presence of a concentration boundary layer at the membrane surface results in prescreening of material before membrane filtration, increasing separation efficiency while decreasing flux. The separation efficiencies of naphthalene with C_{16}-DPPS and C_{12}-DPPS are 98.8 and 94%, respectively. It can be observed from Figure 4.3 that C_{12}-DPDS exhibited almost 10% lower flux than C_{16}-DPDS (for similar surfactant molar concentrations). This is

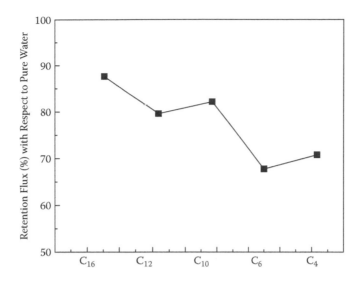

FIGURE 4.3
Relative flux of different chain length diphenyloxide disulfonate surfactant in solution with 10 kDa membrane normalized with distilled water flux.

in agreement with previous studies.[16] The corresponding investigation of disulfonate flux and hydrophobe length indicates that large micelles exhibit higher flux than smaller micelles. Therefore, the process can operate under a degree of concentration polarization for which flux through the boundary layer is limiting. The C_{16}-DPDS surfactant has larger micelles and a higher surface charge density, both leading to its higher aggregation number. The looser packing density resulting from the larger micelle and the greater charge repulsion resulting from the higher surface charge density of this surfactant are both possible explanations for the higher flux observed with increasing alkyl chain length.

At high surfactant concentrations of approximately 50 times the CMC, membrane exclusion has been observed to provide a relatively constant separation with membrane pore size.[17]

4.2 Nonionic Surfactant

Nonionic surfactants consist of hydrophilic head with no charge. Most nonionic surfactants are more tolerant to water hardness, and effective at low concentration since they have good cold water solubility and a low CMC

TABLE 4.1

Characteristics of Triton X Surfactants

Surfactant	PEO Length	Molecular Weight (g/mol)	Molecular Structure
Triton Series of Surfactants			
X-114	8	552.29	
X-100	10	647	
X-305	30	1,527.87	
X-405	40	1,969.40	
Neodol Series of Surfactants			
91-5E	5	380	
91-6E	6	424	
91-8E	8	512	

value.[18] However, the use of nonionic surfactants alone via MEUF is suitable only for removal of organic contaminants.

Among the nonionics of PEO type, Triton X can be customized to have different PEO units. Table 4.1 lists possible Triton X combinations.

Shin et al.[19] have studied the desorption of Cd^{2+} on soil using ligand-modified surfactant systems. Figure 4.4 shows that there exists a linear

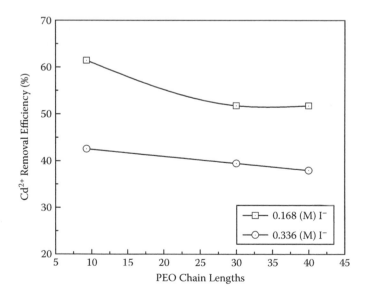

FIGURE 4.4

Effect of PEO chain lengths on the removal efficiency of Cd^{2+} at two different iodide concentrations; surfactant concentration is 0.024 (M).

relationship between the Cd^{2+} removal and the chain lengths. However, with Triton X-100 ($n = 9.5$), Triton X-305 ($n = 30$), and Triton X-405 ($n = 40$), surfactant alkyl chain length affects Cd^{2+} desorption adversely, indicating that metal-ligand complexes are preferably stabilized by the micelles' hydrophobic octyl phenyl (OP) groups rather than by the hydrophilic PEO groups. In another instance of ethoxylated decyl alcohols the solubilization capacity can be ranked as Neodol 91-8E > Neodol 91-6E > Neodol 91-5E. It should be noticed that a decrease in the hydrophilicity, e.g., a shorter polyoxyethylene chain, causes an increase in the aggregation number of the micelle.[20] A comparison of the effectiveness between Triton X-114 and Neodol 91-8E in solubilization can be made because these surfactants have the same number of ethyleneoxide units (n = 8, same hydrophilicity), and differ in hydrocarbon chain structure and length. The linear hydrocarbon chain of Neodol 91-8E has an average carbon atoms number equal to 10.2, while the hydrophobic part of Triton X-114 consists of 8 carbon atoms plus an aromatic ring equivalent to 3.5 linear carbon atoms,[21] namely, an average carbon atoms number equal to 11.5. The increase in solubilization effectiveness of Neodol 91-8E compared to that of Triton X-114 should be due to the difference in hydrophobicity between the two surfactants.[22]

For a surfactant of polyoxyethyleneglycol alkylether, $H(CH_2)_nO(CH_2CH_2O)_mH$ (hereafter denoted as A_nE_m), the average micelle size increases with an increase in the alkyl chain length (n) of a surfactant, while it decreases with an increase in the ethylene oxide chain length (m) of a surfactant. Some of the possible configurations and properties are listed in Table 4.2. It must be noted that the CMC values of the surfactants increase 10 times per decrease of two methylene groups in the hydrophobic part of a surfactant, whereas a change from 8 to 5 ethylene oxide units in the hydrophilic part of a surfactant results only in a slight increase in CMC (Chapter 3).

TABLE 4.2

Characteristics of Different Types of Polyoxyethyleneglycol Alkylether

Surfactant	Molecular Weight (g/gmol)	Micelle Diameter (nm) (determined by dynamic light scattering)
$A_{10}E_8$	510	5.0
$A_{12}E_8$	538	7.0
$A_{12}E_5$	406	25.0
$A_{16}E_8$	594	12.0
$A_{12}E_7$	494	—
$A_{12}E_6$	450	—
$A_{14}E_8$	566	—

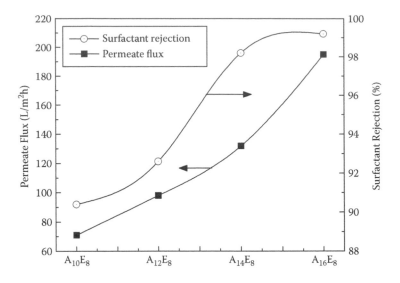

FIGURE 4.5
Variation of permeate flux and surfactant rejection of different alkyl chain surfactant (100 CMC concentration) when ultrafiltered at 300 kPa TMP and 3 kDa CA membrane.

The extent of adsorption decreases with an increase of the numbers (m) of the ethylene oxide group and increases with an increase of the alkyl chain length (n). Interaction forces increase with an increase of the numbers of alkyl chains, while interaction forces slightly decrease with an increase of the numbers of the ethylene oxide group. The molecular weight of the surfactant molecule plays an important role. The longer the chain length, the higher is the size and molecular weight of the surfactant. Thus, micelles with a higher molecular weight surfactant monomer are retained easily by the membrane with a higher retention of the contaminant (refer to Figure 4.5). Since the molecular weight of $A_{16}E_8$ is larger than those of the other surfactants, and the hydrophobic interactions of $A_{16}E_8$ with the polysulfone membrane are stronger than any other cases, the highest surfactant rejection and the most remarkable flux decrease are observed with $A_{16}E_8$ at the same operating pressure. The rejection of the $A_{12}E_5$ surfactant is the lowest at constant n even though the surfactant concentration of $A_{12}E_5$ is the lowest and the average micelle size is the largest among them. It seems that the chain length has a significant effect when the strength of the hydrophobic interaction is slightly changed by the variation of the chain length of the ethylene oxide group. Surfactant rejection can also be affected by the polymer composition of the membrane matrix.[23]

A contrast of ionic and nonionic surfactant behaviors can be observed in Figure 4.6. It is seen from the figure that the solubility of phenol in water

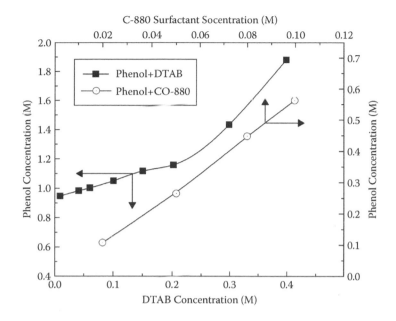

FIGURE 4.6

Solubility diagram of phenol in dodecyltrimethylammonium bromide (DTAB) and nonylphenol ethoxylate having 30 ethylene oxide units (Igepal CO-880) in micellar solutions.

decreases with the addition of a nonionic surfactant; since phenol binds to the ethylene oxide chain,[24] it reduces the interaction between water and the ethylene oxide, thereby reducing surfactant solubility. Thus, increasing the number of units of ethylene oxide increases the number of binding sites that are available at a constant surfactant concentration. To increase the amount of bound phenol using ionic surfactants, the hydrophile and the lipophile can be changed to increase the contribution of the electrostatic free energy associated with the formation of micelles.[25] There is, however, a limited benefit of doing this since any variation of the surfactant structure that increases the electrostatic free energy will generally increase the surfactant monomer concentration. In case of nonionic surfactant, increasing the surfactant lipophile will also decrease the cloud point and result in a smaller single-phase region. Decreasing the lipophile might initially be beneficial by decreasing the free ions, but it also increases the tendency for micelles to penetrate the membrane.[26]

The important considerations described in this chapter are briefly summarized in Figure 4.7. However, during ultrafiltration of a micellar solution, apart from the surfactant, the selection of a suitable membrane is also an important factor in the overall efficiency of the process.

Surfactant Structural Considerations

Ionic: $H_{2m+1}C_mNC_nH_{2n+1}$ with C_nH_{2n+1} above and C_nH_{2n+1} below

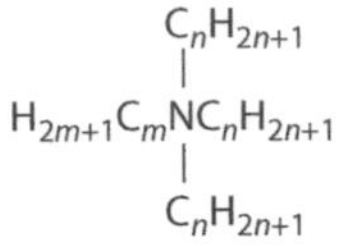

Change	Benefit	Limitation
Increase m	Increase rejection and decrease cmc	Raises Kraft temperature
Increase n	Increase rejection	Increase α_p and cmc
Add inorganic salt	Increase rejection and decrease cmc	Increase α_p
Increase surfactant concentration	Decrease α_p	Critical "leakage" concentration

Nonionic: $H_{2m+1}C_m - O(C_2H_4O)_nH$

Change	Benefit	Limitation
Increase m	Increase rejection and decrease cmc	Decrease cloudpoint temperature
Increase n	None for $n > 15$ to 20	None for $n > 15$ to 20
Increase surfactant concentration	Decrease α_p	Critical "leakage" concentration

FIGURE 4.7
Surfactant structure considerations and its effects on MEUF performance[14] (solute affinity—a_p). (Reproduced with permission from Taylor and Francis Ltd.)

References

1. Scamehorn, J.F., and Harwell, J.H. 1989. *Surfactant-based separation processes.* New York: Marcel Dekker.
2. Misra, S.K., Mahatele, A.K., Tripathi, S.C., and Dakshinamoorthy, A. 2009. Studies on the simultaneous removal of dissolved DBP and TBP as well as uranyl ions from aqueous solutions by using micellar-enhanced ultrafiltration technique. *Hydrometallurgy* 96: 47–51.
3. Yenphan, P., Chanachai, A., and Jiraratananon, R. 2010. Experimental study on micellar-enhanced ultrafiltration (MEUF) of aqueous solution and wastewater containing lead ion with mixed surfactants. *Desalination* 253: 30–37.
4. Huang, J.-H., Zeng, G.-M., Fang, Y.-Y., Qu, Y.-H., and Li, X. 2009. Removal of cadmium ions using micellar-enhanced ultrafiltration with mixed anionic-nonionic surfactants. *J. Membr. Sci.* 326: 303–309.
5. Luo, F., Zeng, G.-M., Huang, J.-H., et al. 2010. Effect of groups difference in surfactant on solubilization of aqueous phenol using MEUF. *J. Hazard. Mater.* 173: 455–461.

6. McBain, J.W., and Richards, P.H. 1946. Solubilization of insoluble organic liquids by detergents. *Ind. Eng. Chem.* 38: 642–646.

7. Saito, H., and Shinoda, K. 1967. The solubilization of hydrocarbons in aqueous solutions of nonionic surfactants. *J. Colloid Interf. Sci.* 24: 10–15.

8. Tokiwa, F. 1968. Solubilization behavior of sodium dodecylpolyoxyethylene sulfates in relation to their polyoxyethylene chain lengths. *J. Phys. Chem.* 72: 1214–1217.

9. Klevens, H.B. 1950. Effect of electrolytes upon the solubilization of hydrocarbons and polar compounds. *J. Am. Chem. Soc.* 72: 3780–3785.

10. Schott, H. 1967. Solubilization of a water-insoluble dye. II. *J. Phys. Chem.* 71: 3611–3617.

11. Rosen, M.J. 2004. *Surfactants and Interfacial phenomena.* Hoboken, NJ: John Wiley & Sons.

12. Dunn Jr., R.O., Scamehorn, J.F., and Christian, S.D. 1985. Use of micellar-enhanced ultrafiltration to remove dissolved organics from aqueous stream. *Sep. Sci. Technol.* 20: 257–284.

13. Boethling, R.S. 1984. *Environmental fate and toxicity in wastewater treatment of quaternary ammonium surfactants.* Washington, DC: EPA.

14. Kandori, K., and Schechter, R.S. 1990. Selection of surfactants for MEUF, separation science and technology. *Sep. Sci. Technol.* 25(1–2): 83–108.

15. Zaghbani, N., Hafiane, A., and Dhahbi, M. 2009. Removal of Eriochrome Blue Black R from wastewater using micellar-enhanced ultrafiltration. *J. Hazard. Mater.* 168: 1417–1421.

16. Roberts, B.L., Scamehorn, J.F., Christian, S.D., Tucker, E.E., and Uchiyama, H. 1992. Phase 1 SBIR Report prepared for the U.S. Air Force, Tyndall Air Force Base. Norman, OK: Surfactant Associates.

17. Dunn Jr., R.O., Scamehorn, J.F., and Christian, S.D. 1987. Concentration polarization effects in the use of MEUF to remove dissolved organic pollutants from waste water. *Sep. Sci. Technol.* 22(2–3): 763.

18. Schmitt, T.M. 1992. *Analysis of surfactants.* New York: Marcel Dekker.

19. Shin, M., Barrington, S.F., Marshall, W.D., Kim, J.-W. 2005. Effect of surfactant alkyl chain length on soil cadmium desorption using surfactant/ligand systems. *Chemosphere* 58: 735–742.

20. Mackay, R.A. 1997. Solubilization. In *Nonionic surfactants: Physical chemistry,* ed. M.J. Schick, 308–314. Surfactant Science Series, vol. 23. New York: Marcel Dekker.

21. Rosen, M.J. 1989. Solubilization by solutions of surfactants: Micellar catalysis. In *Surfactants and interfacial phenomena,* 170–196. New York: John Wiley & Sons.

22. Xiarchos, I., and Doulia, D. 2006. Effect of nonionic surfactants on the solubilization of alachlor. *J. Hazard. Mater.* B136: 882–888.

23. Kim, C.K., Kim, S.S., Kim, D.W., Lim, J.C., and Kim, J.J. 1998. Removal of aromatic compounds in the aqueous solution via micellar enhanced ultrafiltration. Part 1. Behavior of nonionic surfactants. *J. Membr. Sci.* 147: 13–22.

24. Kandori, K., McGreevy, R.J., and Schechter, R.S. 1989. Solubilization of phenol in polyethoxylated nonionic micelles. *J. Colloid Interf. Sci.* 132: 395–402.

25. Larsen, J.W., and Tepley, L.B. 1974. Effect of aqueous alcoholic solvents on counterion-binding to CTAB micelles. *J. Colloid Interf. Sci.* 49: 113–118.

26. Osborne-Lee, I.W., Schechter, R.S., Wade, W.H., and Barakat, Y. 1985. A new theory and new results for mixed nonionic–anionic micelles. *J. Colloid Interf. Sci.* 108: 60–74.

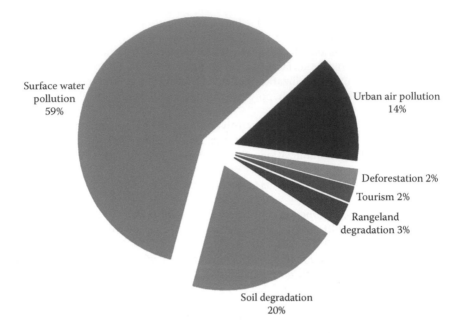

COLOR FIGURE 1.1
Proportion of cost (total cost U.S.$9.7 billion) due to pollution in India.[2]

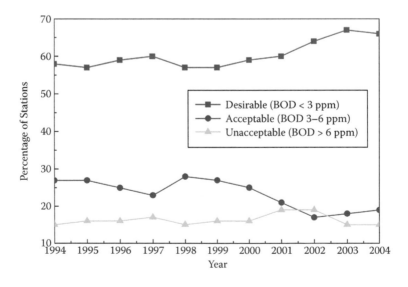

COLOR FIGURE 1.3
BOD of industrial effluent in India.

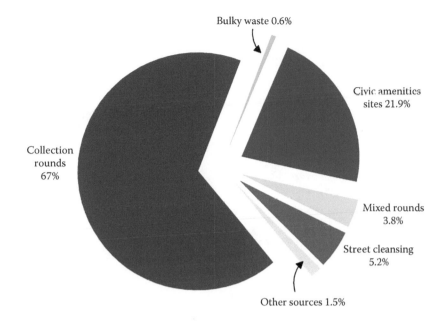

COLOR FIGURE 1.4
Estimated total household wastes for England and Wales in 1993–1994 (22.67×10^6 kg/yr) by main outlets.[8]

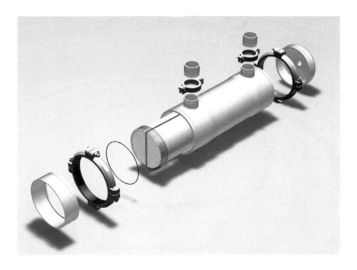

COLOR FIGURE 2.4(b)
Pictures of tubular membrane module.

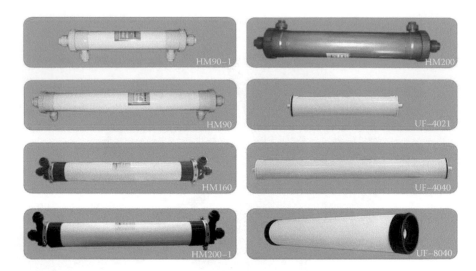

COLOR FIGURE 2.4(c)
Pictures of hollow fiber modules with end cap.

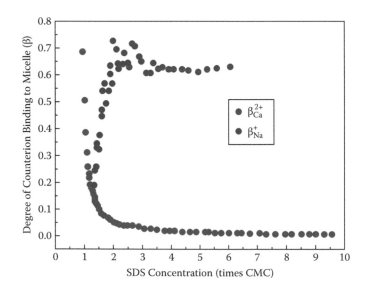

COLOR FIGURE 3.8
Competitive counterion binding for 0.3 mM of $CaCl_2$ solution at 30°C.

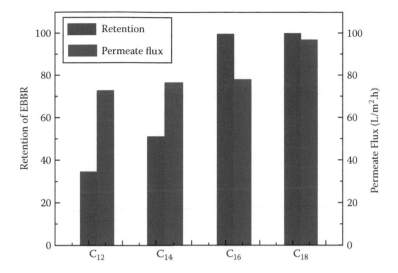

COLOR FIGURE 4.2
Comparison of permeate flux and retention using different alkyl chain lengths of trimethyl-ammonium bromide for removal of 0.001 (M) EBBR dye at 10 CMC surfactant concentration.

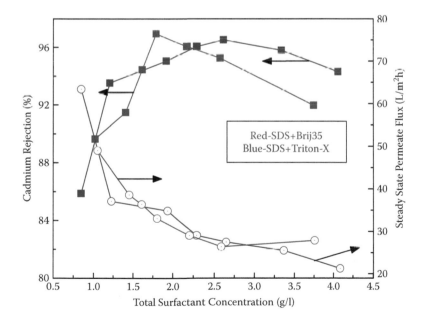

COLOR FIGURE 5.3
Comparison of the permeate flux and rejection using mixed surfactant; SDS concentration in the mixture is 0.86 g/l (operating conditions: TMP 70 kPa, HF membrane 6 kDa, Cd^{2+} concentration 50 ppm).

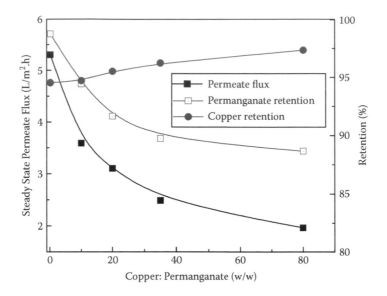

COLOR FIGURE 5.12
Removal of copper and permanganate using 25 kg/m³ SDS and 10 kg/m³ CPC; 5 kDa TFC membrane, TMP 414 kPa, and 30 LPH.

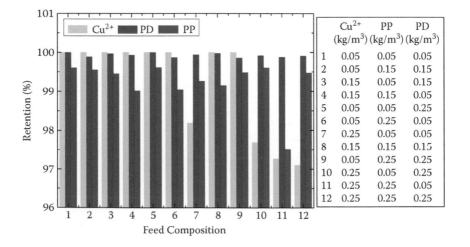

COLOR FIGURE 5.15
Effect of feed composition on retention of solutes; TMP 414 kPa, 5 kDa membrane, and cross-flow rate 60 LPH.

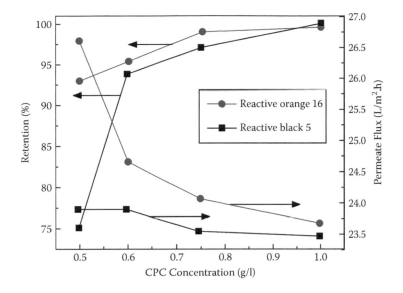

COLOR FIGURE 6.3
Variation of rejection and permeate flux at different CPC concentrations for MEUF of reactive dyes using 10 kDa membrane, TMP 300 kPa, and feed concentration 50 ppm.

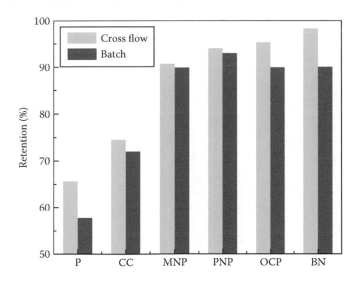

COLOR FIGURE 6.5
Relative retention of different phenols using 10 kg/m³ CPC and 0.24 kg/m³ solute concentration; TMP 345 kPa, 1 kDa membrane and concentration, TMP 345 kPa, 1 kDa membrane, and 30 LPH cross-flow rate.

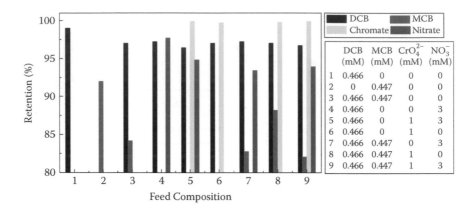

COLOR FIGURE 6.12
Retention of different solutes using 5 kDa PS membrane in cross-flow mode using 20 mM CPC at 200 kPa.

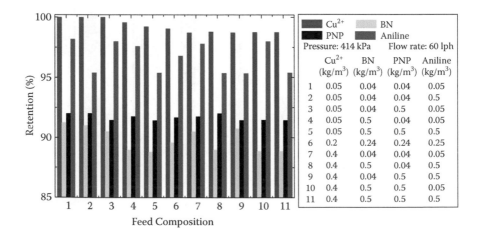

COLOR FIGURE 6.14
Effect of feed composition on solute retention in Cu, BN, PNP, and aniline mixture (5 kDa PS membrane).

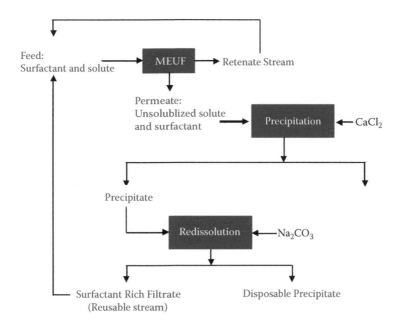

COLOR FIGURE 8.1
Schematic of a typical anionic surfactant recovery process along with MEUF.

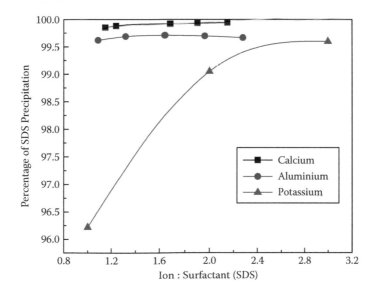

COLOR FIGURE 8.2
Variation of precipitation of SDS with changing ion/surfactant ratio.

5

Removal of Inorganic Pollutants

Many industrial wastewater streams contain a potential amount of heavy metals that is far above the environmental norms of effluent discharge. The metal ions are nonbiodegradable, highly toxic, and may have a significant carcinogenic effect.[1] Due to their high solubility in the aquatic phases, heavy metals can be absorbed by living organisms and accumulate in human organs.[2,3] Table 5.1 lists some of the possible health effects of the presence of inorganic ions in the universal solvent water.

Nowadays, the economic operation of the membrane processes draws attention to achieving lower costs, in practice. Employment of lower trans-membrane pressure of the selected membrane application is important with regard to high flux achievement, as well as obtaining the permeate of high quality. It can be emphasized that micellar-enhanced ultrafiltration (MEUF) is one of the separation technologies that can be effectively used for the removal of almost all cations and anionic with relatively high efficiency, enough for safely discharging into the environment. One of the added advantages of MEUF over conventional metal removal processes such as chemical precipitation, adsorption, ion exchange, chelating, evaporation, coagulation-flocculation, flotation, electrodeposition, liquid-liquid extraction, etc., is that it is free from secondary pollution of disposal of the retentate, since the pollutants are solubilized in the surfactant micelles, and do not have environmental and health hazards.

5.1 Single Component System

Trace amounts of copper(II) ions can be effectively removed from wastewater streams using MEUF. Using an anionic surfactant, SDS is very effective compared to other anionic surfactants. Li et al.[4] have reported that copper rejection is around 92% for an SDS concentration of 10 mM and more, using a 10 kDa UF membrane. However, adding a nonionic surfactant (Triton X), for increasing the efficiency of SDS, can increase the rejection up to 99% (SDS concentration 10 mM and Triton X 3.0 mM)[4] at a copper concentration of 0.2 mM and pH 5. The effects of stirring, transmembrane pressure (TMP), and pH on rejection of copper were studied by Juang et al.[5] It has been found that maximum rejection increases slightly from 88% to 94%, when TMP has

TABLE 5.1

Health Effects Caused by the Presence of Inorganic Ions

Contaminant	Potential Health and Other Effects
Aluminum	Can precipitate out of water after treatment, causing increased turbidity or discolored water. Lasting uptakes of significant concentration of aluminum can lead to damage of central nervous system, loss of memory, dementia, and listlessness.
Arsenic	Causes acute and chronic toxicity, liver and kidney damage, neurological disorders; decreases blood hemoglobin. Possible carcinogen.
Cadmium	Replaces zinc biochemically in the body and causes high blood pressure, liver and kidney damage, and anemia. Destroys testicular tissue and red blood cells. Toxic to aquatic biota.
Copper	Can cause stomach and intestinal distress, liver and kidney damage, anemia in high doses. Imparts an adverse taste and significant staining to clothes and fixtures. Essential trace element but toxic to plants and algae at moderate levels.
Chromium	Chromium III is a nutritionally essential element. Chromium VI is much more toxic than chromium III and causes liver and kidney damage, internal hemorrhaging, respiratory damage, dermatitis, and ulcers on the skin at high concentrations.
Iron	Imparts a bitter astringent taste to water and a brownish color to laundered clothing and plumbing fixtures.
Lead	Affects red blood cell chemistry; delays normal physical and mental development in babies and young children. Causes slight deficits in attention span, hearing, and learning in children. Can cause slight increase in blood pressure in some adults. Carcinogenic.
Manganese	Causes aesthetic and economic damage, and imparts brownish stains to laundry. Affects taste of water, and causes dark brown or black stains on plumbing fixtures. Relatively nontoxic to animals but toxic to plants at high levels.
Mercury	Causes acute and chronic toxicity. Targets the kidneys and can cause nervous system disorders.
Zinc	Aids in the healing of wounds. Causes no ill health effects except in very high doses. Imparts an undesirable taste to water. Toxic to plants at high levels.
Nickel	Damages the heart and liver of laboratory animals exposed to large amounts over their lifetime.
Fluoride	Decreases incidence of tooth decay but high levels can stain or mottle teeth. Causes crippling bone disorder (calcification of the bones and joints) at very high levels.
Chloride	Deteriorates plumbing, water heaters, and municipal waterworks equipment at high levels. Above secondary maximum contaminant level, taste becomes noticeable.
Nitrate	Toxicity results from the body's natural breakdown of nitrate to nitrite. Causes "blue baby disease," or methemoglobinemia, which threatens oxygen-carrying capacity of the blood.
Dissolved solids	May have an influence on the acceptability of water in general. May be indicative of the presence of excess concentrations of specific substances not included in the Safe Water Drinking Act, which would make water objectionable. High concentrations of dissolved solids shorten the life of hot water heaters.

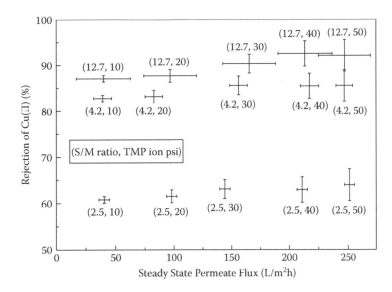

FIGURE 5.1
Representation of rejection and steady-state permeate flux during MEUF using SDS at different operating conditions (surfactant to metal ratio and TMP) at pH 4.

been increased from 60 to 350 kPa at pH 5, stirrer speed 300 rpm, and surfactant-to-metal ratio 12.7. However, the steady-state flux is maximum at pH 3 within the observed TMP range. Flux increases linearly on increasing TMP with SDS. The flux decline mechanism shows strong evidence of complete and intermediate blocking in the initial stages of ultrafiltration. This phenomenon is more prominent using SDBS rather than SDS. The size of the micelle plays an important role in the blocking process. A novel technique combining electrolysis and MEUF increases the rejection efficiency as well as decreases the hydraulic retention time by four times.[6] Figure 5.1 shows the dependence of the operating criterion for enhanced copper rejection and permeate flux.

One of the toxic divalent ions significantly present in the wastewater stream is cadmium ions. Large-scale treatment of effluent water has been quite popular based on the MEUF technique. Removal of cadmium ions using MEUF has a rejection efficiency of 85 to 99% using SDS as the surfactant.[7] The removal efficiency is sensitive to pH and to the presence of electrolyte.[8] The rejection efficiency significantly improves from 84% to 98% on increasing the pH from 2.5 to 9 using a hollow fiber membrane (TMP 70 kPa) at an SDS concentration of 8.0 mM. The rejection at a lower pH is less because of the fact that there exists a competitive binding between H^+ and Cd^{2+} to the micelle. With H^+ being higher in concentration at lower pH, and also because of its smaller size, it is preferentially bound to the micelle. The rejection efficiency increases on increasing the pH of the solution that reaches the asymptotic limit at pH 9. Rejection of cations is more favored at a pH greater than 7 since OH⁻ enhances the transport of the Cd^{2+} to the micelle for effective binding. Similarly, on adding a strong electrolyte like NaCl, the rejection efficiency decreases considerably. This is due to the reason that the transport of Na^+ is more preferred than Cd^{2+}, as Na^+ has lower values of Debye length owing to is lower valency. The effect on the rejection efficiency due to the presence of a cosurfactant has been analyzed by Fang et al.[9] It has been shown by Fang et al.[9] that a mixed micelle system of SDS + Brij 35 has a higher rejection efficiency than SDS + Triton X up to a nonionic-to-ionic surfactant ratio of 0.5. Beyond that, both the systems have comparable rejections when ultrafiltered using a 10 kDa molecular weight cutoff (MWCO) membrane operated at 70 kPa and neutral pH. However, the permeate flux is higher for the SDS + Triton X system for the same operating conditions. A response surface-based methodology has also been developed to improve the process parameters' interactions and optimization.[10] The effects of surfactant concentration using only anionic surfactant (SDS) and mixed surfactant, on the retention and permeate flux, have been represented in Figures 5.2 and 5.3, respectively.

Hexavalent chromium can be effectively removed by use of MEUF.[11] It has been observed that using a 10 kDa MWCO membrane and CPC as the surfactant, the retention is higher using a higher surfactant concentration, in the range of 0.1 to 60 mM. As the CPC concentration increases from 10 to 60 mM at TMP 580 kPa, the permeate flux decreases considerably from 9.9 to 6.7 $L/m^2.h$ on increasing CPC concentration from 10 to 60 mM at TMP 580 kPa due to the

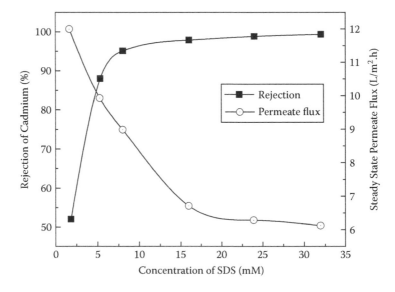

FIGURE 5.2
Rejection and permeate flux for removal of cadmium with varying SDS concentrations using HF membrane 6 kDa, TMP 70 kPa, and feed Cd^{2+} concentration 100 ppm.

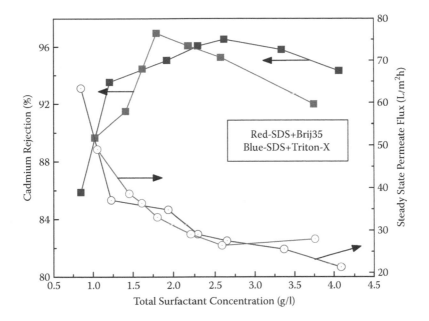

FIGURE 5.3 (See color insert.)
Comparison of the permeate flux and rejection using mixed surfactant; SDS concentration in the mixture is 0.86 g/l (operating conditions: TMP 70 kPa, HF membrane 6 kDa, Cd^{2+} concentration 50 ppm).

increasing resistance of the micellar aggregation layer thickness. However, increasing the CPC concentration increases the retention of the metal ions up to 99% for a CPC beyond 30 mM. Hence, it can be concluded that the optimum CPC concentration is between 10 and 20 mM. Figure 5.4 illustrates the variation of cadmium retention and permeate flux with CPC.

Addition of salt to the system is particularly useful in reducing the CMC. It has been studied by Gzara et al.[12] It is observed that CPC permeate concentration decreases from 1 to 0.15 mM, when the concentration of NaCl increases from 1 to 500 mM, whereas chromate rejection remains higher than 90%, even in the presence of 100 mM of NaCl. Beyond this concentration, rejection decreases and reaches 46% for 0.5 M NaCl. Using CTAB as the surfactant, it has been shown by Gzara et al.[12] that maximum rejection can be obtained up to 85% in the studied surfactant concentration range of 0.001 to 80 mM. The optimum concentration of CTAB is 0.4 to 0.7 mM at a TMP of 3 bar considering the permeate flux (210 to 230 L/m².h) and rejection (~88 to 93%). Aoudia et al.[13] has analyzed the retention characteristics of Cr^{3+} using SDS and nonyl phenol ethoxylate (NPE) as the mixed surfactant system. It has been found that for a transmembrane pressure (TMP) of 250 kPa, 80% SDS and 20% NPE yield the optimum results when used above the CMC values (99.8% retention compared to 98.5% when using only SDS). The retention is

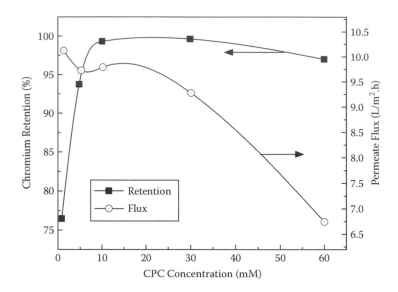

FIGURE 5.4
Effect of surfactant concentration on retention of 50 ppm[Cr(VI)] using 30 mM CPC, 580 kPa TMP, and 10 kDa membrane.

maximum when the total surfactant concentration (0.8 SDS and 0.2 NPE) is greater than 3 mM. Also, as the feed concentration of chromium is increased, the permeate concentration increases linearly, thereby decreasing the retention. Similar to rejection of cadmium, on addition of electrolyte (NaCl), the rejection efficiency decreases using an anionic micelle.

MEUF of Ni^{2+} has been studied by several researchers in recent times.[14-16] An optimum surfactant-to-metal ratio has been obtained by Chhatre and Marathe[14] for a dead-end system using a 20 kDa membrane and SDS as the surfactant. The optimum ratio has been observed to be 10 corresponding to a maximum rejection of 99.5% at an operating pressure of 400 kPa and neutral pH. Similar to the removal of other multivalent ions discussed above, the rejection of Ni^{2+} is strongly influenced by changing pH and addition of salt to the system. For a pH above 3.5, the rejection efficiency is observed to be more than 99%. The presence of salt decreases the rejection, which is more significant for NaCl than NaBr and NaI, when used at a higher concentration (above 10 mM). Yurlova et al.[15] have reported an optimum operating TMP of 400 kPa corresponding to a rejection of 83% and permeate flux of 88.6 L/m².h using an SDS concentration of 2.34 mM. The mean pore size of the membrane used in the study is 50 to 200 nm. In another study by Danis and Aydiner,[16] the rejection characteristics have been analyzed using sodium lauryl ether sulfate (SLES) as the surfactant. They have reported that using a 0.2 μm pore size membrane at neutral pH in cross-flow mode (cross-velocity 6 m/s), a maximum observed retention of 98.6%, corresponding to a surfactant-to-metal ratio of 10, is achieved. The steady-state permeate flux is 302.4 L/m^2.h (TMP of 250 kPa), which does not seem to vary with the added surfactant to the system above the CMC. On increasing the TMP, the steady-state flux increases without affecting the retention much in the range of 150 to 250 kPa. Also, adding NaCl to the system at a concentration of 22 mM can decrease the observed rejection to 42% and reduce the permeate flux. Since on addition of electrolyte the micelle size decreases, the fouling mechanisms like complete pore blocking[17] and partial blocking[18] are more prominent and responsible for a decline of flux in the transient state. The change of rejection and permeate flux with SDS concentration is shown in Figure 5.5a. On the other hand, Figure 5.5b presents the optimum ratio of the mixed surfactant in the MEUF for removal of Ni(II).

Zinc at low and high concentrations can be effectively removed using MEUF.[19] A 20 kDa membrane is used in cross-flow and continuous mode without recycling back the permeate or retentate. The rejection efficiency approaches the asymptotic limit on increasing the SDS concentration (refer to Figure 5.6). Typically, retention of more than 95% is obtained when the surfactant concentration is more than 4 mM at an operating pressure of 2 bar. However, there is no significant influence of operating pressure on the rejection efficiency. Also, the rejection efficiency increases from 75% to 98% and permeate flux decreases from 185 to 160 L/m^2.h on increasing the pH from 2 to 12. Thus, from an economical point of view, the process should be operated at higher TMP and higher pH. In another study by Zhang et al.[20] the removal of Zn^{2+} at low surfactant concentration (0.2–3 times CMC) has been analyzed. A rejection of 98.5% is obtained when operated below the CMC and high operating pressure due to the development of a concentration polarization layer over the membrane surface.

An optimum rejection efficiency has been obtained by Huang et al. using a 10 kDa membrane corresponding to a surfactant (SDS)-to-metal ion ratio of 8:10 at an initial Zn^{2+} concentration of 50 ppm and TMP 70 kPa.[21] The rejection of Zn^{2+} increases with the ratio of surfactant-to-metal ratio S/M. When the S/M ratio is less than or equal to 5.8, Zn^{2+} rejection increases by 13.8% with a doubled S/M ratio. As the S/M ratio is higher than 5.8, the rejection of Zn^{2+} increases proportionately. Moreover, when the S/M ratio is higher than 24.4, the rejection of Zn^{2+} reaches 97%. The rejection reaches above 90% when the S/M ratio is above 8 and 97% when above 24.4 mM.

One of the noble metals like gold can also be removed from waste streams in trace amounts using MEUF.[22] Polyoxyethylene nonyl phenyl ethers (PONPEs) with an average ethylene oxide (EO) number of 10 have been used to analyze the retention characteristics of gold. A 3 kDa membrane has been operated under an operating pressure in the range of 50 to 300 kPa in a stirred batch cell at 200 rpm (metal ion concentration 0.01 M). The results clearly show that when the surfactant concentration is greater than 0.01 M, the rejection is definitely greater than 90%, at operating pressure of 100 kPa. Also, on increasing the EO number from 10 to 20, the rejection increases

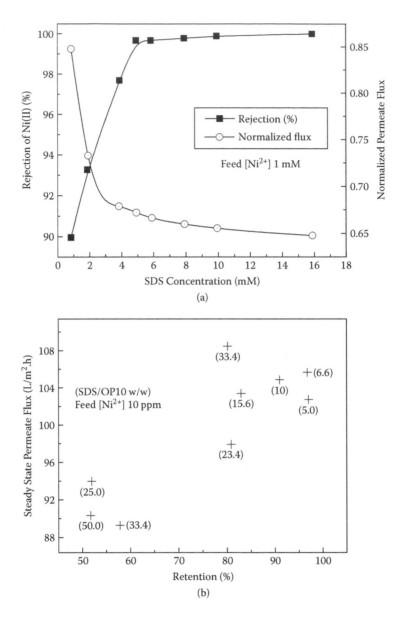

FIGURE 5.5
Effect of surfactant concentration on the retention and permeate flux for MEUF of Ni^{2+} using 20 kDa membrane at 400 kPa TMP: (a) only SDS, (b) using a mixture of SDS and monoalkylphenol polyethoxylate (nonionic).

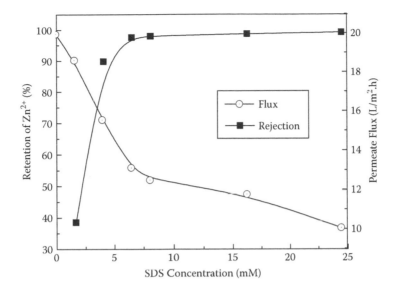

FIGURE 5.6
Effect of SDS concentration on the removal of Zn^{2+} (50 ppm), TMP 70 kPa, 6 kDa membrane.

from 91.6 to 95% at a surfactant concentration 0.01 M and 100 kPa TMP, but the flux is apparently constant at 11.4 $L/m^2.h$. The EO number signifies that the electron-donating groups of the surfactant have more affinity toward gold(III), and hence it is more advantageous to use a surfactant that has more EO groups. The effect of operating pressure is similar to the removal of previous ions; the retention is almost constant at 91% (surfactant concentration 0.01 M), but the flux changes from 11.4 to 24 $L/m^2.h$ on changing the TMP from 100 to 300 kPa. One interesting observation is that in the presence of dilute HCl (which is necessary to dissolve noble metals like gold, platinum, palladium, etc.) the noble metals exist in anionic form but base metals like zinc, aluminum, copper, iron, etc., take cationic forms.[23] This fact is evident from the negative values in Table 5.2, which signifies that the noble metals do not bind to the anionic surfactants. Table 5.2 also signifies that PONPE is highly selective, particularly toward retention of gold. Moreover, PONPE

TABLE 5.2

Relative Comparison of Removal Efficiency Using Different Types of Surfactants

Surfactant	Surfactant Concentration (M)	Flow Rate (m/s)	Retention (%)				
			Gold	Platinum	Copper	Iron	Zinc
CPC	0.01	0.011	100.0	99.9	−13.2	−11.1	−12.8
SDS	0.01	0.011	−10.2	−4.6	85.3	98.2	84.8
PONPE	0.01	0.011	89.0	2.4	5.1	8.9	8.7

TABLE 5.3

Different Removal Efficiencies of Pb^{2+} Using Different Concentrations of Non-ionic Surfactants and SDS

Nonionic Surfactant (mole fraction)	0.1		0.2		0.3	
SDS Concentration (mM)	TX-100	NP12	TX-100	NP12	TX-100	NP12
4.2	0.9654	0.9677	0.9551	0.9624	0.9312	0.9330
8.2	0.9757	0.9803	0.9679	0.9705	0.9524	0.9540
12.3	0.9847	0.9870	0.9771	0.9805	0.9632	0.9643

has much less surfactant leakage during MEUF, which is also an important consideration in this regard.

Removal of lead from wastewater has been studied by Yenphan et al. using a mixed surfactant system.[24] The rejection of Pb^{2+} has been analyzed using an anionic surfactant (SDS) and nonionic surfactant, namely, TX-100 and NP12. The ultrafiltration is performed using a polyether sulfones (PES) membrane of MWCO 10 kDa in batch mode at a TMP of 10 kPa. The feed concentration is fixed at 20 ppm, which is equivalent to the toxic content of wastewater discharge from a battery plant. It must be realized that the rejection efficiency of a nonionic surfactant is very poor, only 38% for NP12 and 17% for TX-100. Hence, it can only act as an enhancer agent to SDS when used at optimum concentration. The number of micelles increases with nonionic surfactant concentration, which enhances the rejection of surfactants and Pb^{2+} as well. However, when the mole fraction of nonionic surfactant in the micelles further increases, the micelle charge density decreases and the counterion binding capacity is reduced.[9] Table 5.3 shows the rejection of Pb^{2+} at different mole fractions of the two nonionic surfactants. The optimum mole fraction is 0.1 and the SDS concentration is 12.3 M for both TX-100 and NP12. The rejection decreases as one increases the mole fraction of the nonionic surfactant (Table 5.3).

At an SDS concentration of 4.2 mM, the permeate flux decreases from 120 to 71 L/m².h as one increases the mole fraction from 0.1 to 0.3 for both nonionic surfactants. However, the rejection of the surfactant increases as one increases the mole fraction of the nonionic surfactant. Hence, operating below 0.1 M fraction is not viable considering decreased surfactant rejection.

The removal of another precious metal, palladium, has been studied by Ghezzi et al.[25] The experiment was performed using a 3 kDa Millipore membrane, operated at 300 kPa and feed Pd^{2+} concentration 8.5×10^{-5} (M). A cationic surfactant, namely, dodecyl-trimethylammonium chloride (DTAC), is used. It has been observed that the rejection efficiency of more than 95% is achieved at pH 3.5 when the DTAC concentration is higher than CMC. It must be noted that the pH of the solution has a significant effect on the equilibrium of the $PdCl_3^-/PdCl_4^{2-}$ and $Pd(OH)_2$. The shifts in equilibrium due to the micellar solution are attributed to the changes in activity coefficients of the ions[26] and proton association/dissociation on the micelle surface.[27] The presence of DTAC alters the onset of precipitation of $Pd(OH)_2$ that occurs at pH 4 in aqueous solution.[28] It can be concluded that the presence of the micelle favors the formation of $PdCl_3^-$ compared to $Pd(OH)_2$, which is manifested in the high rejection at pH around 3.5 and DTAC concentration 0.005 (M), and precipitation starts at pH 7 due to the presence of DTAC. Also, removal of another precious metal platinum is also hydrolyzed using HCl. It has been studied by Gwicana et al. that using a PES membrane and CPC, more than 95% removal is achieved.[29] However, it must be realized that simultaneous removal of platinum and palladium is itself difficult due to the instabilities at different pH values.

One of the radiogenic isotopes of strontium is ^{87}Sr, which is naturally present in 7 to 10% of all the isotopes of Sr. Hence removal of Sr, even in trace amounts, is extremely important. MEUF of Sr^{2+} using di-2-ethylhexylphosphoric acid (DEHPA) has been analyzed by Rajec and Paulenova.[30] The solution containing 10^{-5} (M) feed Sr^{2+} is ultrafiltered using 10 kDa membrane and 0.1 (M) DEHPA or dithio-DEHPA. The rejection efficiency increases on increasing the pH of the solution. Ultrafiltration above pH 4 is quantitatively feasible since rejection is above 95% and attains saturation at pH 5. When used above the CMC values, the thio-DEHPA shows higher rejection efficiency than DEHPA up to the studied surfactant concentration of 0.005 (M). At pH 5, the rejection using 0.0025 (M) thio-DEHPA is 92%, while using the same concentration of DEHPA, it is 53%. Use of NaCl has a signification reducing effect on the metal ion removal due to the presence of a competitive binding effect.[31]

The removal of As(V) has been particularly effective when using a cationic surfactant CPC. Iqbal et al.[32] have shown that among four different cationic surfactants, CPC exhibits the highest removal efficiency (96%) of arsenic, followed by CTAB (94%) under the operating conditions of 243 ppb As(V) feed concentration, 10 kDa membrane, pH 7, and 10 mM surfactant concentration. The removal efficiency is reduced at lower pH. The extent of arsenic(V) rejection does not change considerably using a dead-end or cross-flow mode of filtration. Arsenic removal has been investigated using CPC at a lower concentration of 10 mM by Beolchini et al.[33] The authors have studied the effect of main operating conditions and process parameters on the removal efficiency. They have reported that arsenic concentration in the permeate is around 1 ppm for a feed concentration of 10 to 40 ppm at a considerably lower CPC concentration of 1 to 3 ppm and using a large pore size membrane of 50 kDa. This is particularly useful in obtaining high flow rate when extremely contaminated streams of large volumes are to be treated[34] from an economical perspective. One can develop a multistage diafiltration setup by recycling the recovered surfactant to produce permeate concentration below the permissible limits. At 10 ppm arsenic concentration, 90% reduction of As in the permeate is obtained using CPC. However, on increasing surfactant concentration from 1 to about 2.5 mM, the retention does not improve. At 40 ppm arsenic concentration, a partial reduction of As(V) concentration in the permeate is obtained only in the first step using 2 mM CPC concentration, which is not observed in the subsequent stages due to micelle saturation. Removal of arsenic in the range of 0 to 221 ppb using 5 and 10 MWCO PES and regenerated cellulose (RC) membranes has been studied by Gecol et al.[35] The study reveals that the RC membrane provides better removal than PES membranes at low CPC concentration due to its negatively charged surface at the experimental conditions. With the addition of 10 mM CPC, the arsenic concentration is reduced below the permissible limit (<10 ppb as set by the USEPA), irrespective of the feed concentration in both membranes. Highest arsenic removal, 100%, is achieved for the feed water arsenic concentrations of 22 and 43 ppb using 5 kDa PES membranes at pH 5.5 and a 10 kDa RC membrane at pH 8. Regardless of the feed water arsenic concentration (in the range of 20 to 45 ppb feed concentration), 100% arsenic removal is obtained with a 5 kDa PES membrane at pH 8. However, in the presence of other inorganic solutes the removal efficiency of As(V) does not change but the permeate flux reduces in the presence of other co-occurring solutes. It is 56.5 ± 2.5, $43.5 \pm 3.5\%$, $45.5 \pm 7.5\%$, $43.5 \pm 6.5\%$, and $37.5 \pm 6.5\%$ of pure distilled water flux for UF of arsenic and CPC only and in the presence of co-occurring solutes HCO_3^-, HPO_4^-, H_4SiO_4, and SO_4^{2-} species, respectively.[36] Arsenic from contaminated groundwater can also be effectively removed by adsorption in a laterite bed.[37,38] Economic comparison would favor the use of laterite adsorption over MEUF since the cost involved in purchasing CPC is relatively more.

Fluoride contamination in groundwater is one of the major concerns. According to WHO guidelines, the permissible limit of fluoride content for safe drinking water is 1.5 mg/L. The uses of inorganic fertilizers, untreated effluent discharges, and other hydrological activities are the major reasons for increased fluoride content in the underground water and in the environment. Primarily, the calcium rocks (e.g., fluorspar) are responsible for binding fluoride in the soil; however, inorganic fertilizers, on being absorbed in the soil, decrease their retention capacity. Also, fluoride, being highly electronegative, easily binds to other materials. Use of a cationic surfactant, CPC, is very effective in solubilizing fluoride, and thereby removal by micellar-enhanced ultrafiltration. In experimental analysis of a synthetic solution containing 15 ppm (10 times the safe limit) in cross-flow ultrafiltration at 60 LPH and 276 kPa TMP, using a 10 kDa membrane, rejection is above 90% for a CPC concentration of 36 mM and above. Figure 5.7 shows the rejection and permeate flux profile for different feed concentrations of CPC. It can be observed that there exists an optimum concentration of CPC considering a decrease in flux with concentration. For the chosen operating conditions, in this case, the optimum concentration of CPC is 40.5 mM.

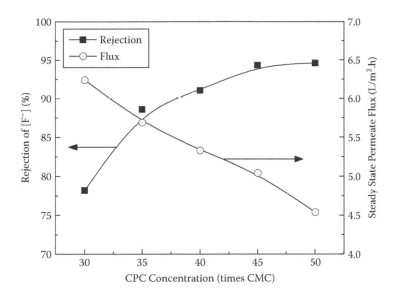

FIGURE 5.7
Retention and permeate flux profile of fluoride removal. Operating conditions: 40 psi, flow rate 60 LPH, membrane 10 kDa; feed concentration of fluoride is 15 ppm, feed volume 4 L, and pH 3.

Uranium is one of the common global contaminants of energy sites (nuclear energy processing and weapon development) and at mine tailing sites associated with its production. Uranium is present as U(IV) and U(VI) in typical subsurface environments. Given the significant health risks of uranium, numerous polluted sites require decontamination. As per the norms of the USEPA, the maximum permissible limit of uranyl ions in drinking water is 30 ppb. To avoid the high costs associated with excavation and disposal of contaminated soils in low-level radioactive waste sites, alternative technologies need to be developed. Furthermore, emphasis on *in situ* remediation schemes is warranted since those technologies are potentially more economical and reduce workers' exposure to contamination. Ligand-modified micellar-enhanced ultrafiltration (LM-MEUF) requires a ligand that consists of a chelating group and a hydrophobic moiety. Such ligands are able to bind a target metal ion and then solubilize in surfactant micelles.[39] High rejection is obtained using an anionic surfactant for removal of cationic solutes; however, high rejection at moderate and high concentrations and sensitive environments requires the use of hydrophobic ligands (chelating agents) for efficient removal.[40]

It is observed from Figure 5.8 that including a strong hydrophobic moiety in the ligand increases the rejection significantly. Also, as the pH increases, the retention increases due to the use of a cationic surfactant. The rejection factor increases with increasing trioctylphosphine oxide (TOPO) concentration up

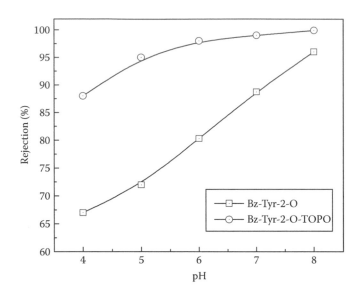

FIGURE 5.8
Effect of pH on retention of uranyl ions using hexadecyl-trimethyl-ammonium bromide (HTAB) 0.04 M and ligands Bz-Tyr-2-O and mixture of Bz-Tyr-2-O + TOPO; ligand concentration 1 mM and UO_2^{2+} 0.05 mM; 300 kPa TMP and 10 kDa CA membrane.

to a saturation limit that probably corresponds to the quantitative formation of the ternary complex. The presence of the highly hydrophobic TOPO molecule in these ternary species could favor the very effective binding to the micellar aggregates. In fact, if the complex is more deeply solubilized in the micelle, the attractive interactions between the aromatic groups of the ligands and the positively charged surfactant head groups can be facilitated.[41] Using the ligand modification of TOPO along with Bz-Tyr-2-O increases the rejection above 99% at normal pH, which can be potentially useful in processing drinking water free of uranyl ions.

At a very low dilute concentration, high removal efficiencies have been obtained by Reiller et al.[42] using only anionic surfactants. Figure 5.9 summarizes the main findings, which shows that rejection is almost constant at 10 CMC surfactant concentration in the studied range. The experiments were conducted in continuous cross-flow mode. The pH of the solution is made highly acidic to avoid hydrolysis of uranyl ions. Use of an anionic surfactant enables the pollutants to be adsorbed on the surface of the micelles; unlike the ligand-modified MEUF, the solutes are bound inside the core of the micelle. Thus, in order to remove uranyl ions one has to select ligand-modified MEUF in case of moderate and higher solute concentration and an anionic surfactant to process the dilute stream. Uranium contamination in soil can be removed by use of a surfactant. Readers should look into the study of Gadelle et al.[43] for a detailed analysis.

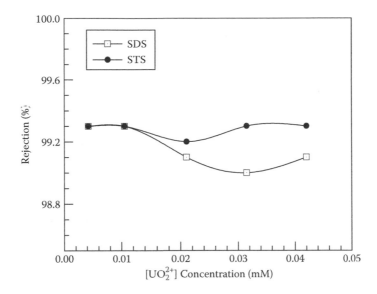

FIGURE 5.9
Removal of U(VI) using two different anionic surfactants—SDS and sodium tetradecyl sulfate (STS); pH 2, surfactant concentration is 10 times that of CMC, TMP 1 bar, flow rate 36 LPH, 10 kDa membrane.

5.2 Multicomponent System

In most industrial effluents, there is evidence of the presence of inorganic and organic pollutants.[44,45] This is one particular area where the MEUF scores significantly higher than the conventional biosorption or activated carbon process because on account of competitive adsorption, the respective removal efficiencies of the individual contaminants are lower, unlike the case of single metal ion removal.[46,47]

5.2.1 Cationic-Cationic Mixture

Simultaneous removal of nickel and zinc has been reported by Channarong et al.[48] using SDS as a surfactant. They have used two different MWCO membranes of 100 and 300 kDa. The maximum average rejection of both metals (equimolar amount of 0.5 mM for each ion) in a mixture is 97.5% for each species (SDS-to-metal ratio is 59.5:1:1) operated at 1 bar and a 100 kDa membrane with a permeate flux of 2.4 LPH. Table 5.4 presents retention values of nickel and zinc for an equimolar feed composition of 0.5 mM, TMP 1.0 bar, and permeate flow rate of 40 ml/min.

The removal efficiency can be increased up to 98% even using a 300 kDa membrane by a MEUF-activated carbon fiber process. When nickel and

TABLE 5.4

Removal Efficiencies Using 100 and 300 kDa Membranes (the values in parentheses correspond to 300 kDa)

SDS:Zn:Ni	17:1:1	25.5:1:1	42.5:1:1	51:1:1	59.5:1:1
Ni removal efficiency in %	92.2	96.3	95.7	96.0	97.5
		(88.6)	(89.3)	(90.7)	(89.0)
Zn removal efficiency in %	91.9	96.7	96.1	96.4	97.5
		(86.5)	(88.7)	(90.1)	(90.0)

cobalt (feed concentration 1 mM) are removed simultaneously, using a 20 kDa membrane, the maximum removal efficiency that can be achieved is 99.5% for both ions at a surfactant (SDS)-to-metal ratio (S/M) of 7.[49] However, at a lower S/M ratio, the removal efficiency of cobalt is higher than that of nickel. The removal efficiency is unaffected by the pH of the system. As the concentration of the contaminants increases, the retention as well as the permeate flux decreases. For example, at an S/M ratio of 7, the retention for cobalt and nickel decreases from 99.5 (both) to 93.8 and 94.5, respectively, and also the flux decreases from 27 to 2.1 L/m².h when the metal ion concentration is increased from 1 mM to 8 mM. However, the decrease in flux can be compensated by increasing the TMP. A theoretical model for prediction of permeate flux considering a gel-controlled filtration for MEUF of binary mixtures of metal ions has been developed by Das et al.[50] Separation of several binary combinations of metal ions has been studied by Juang et al.[51] The experiments are performed in a stirred batch cell using a 1 kDa PA-TFC membrane, at TMP 304 kPa, and using SDS. Table 5.5 shows that separation of metal ions with different valences is more prominent than with similar valences. However, with a more porous membrane, use of a higher S/M ratio increases the removal efficiency. For example, using an S/M ratio of 20 and a 5 kDa MWCO membrane, removal of Zn^{2+} and Cu^{2+} (3.75 mM each) is simultaneously 99.55 and 99.28%, respectively.[52]

TABLE 5.5

Simultaneous Removal of Binary Mixtures of Cations (metal ion concentration 50 ppm, S/M ratio 5)

M_I/M_{II}	$R(M_I)$	$R(M_{II})$
Sr^{2+}/Mn^{2+}	93.8	73.8
Sr^{2+}/Zn^{2+}	97.0	87.5
Zn^{2+}/Co^{2+}	75.2	75.2
Zn^{2+}/Cu^{2+}	83.5	80.2
Cr^{3+}/Cu^{2+}	93.1	60.1
Cr^{3+}/Zn^{2+}	92.4	42.8

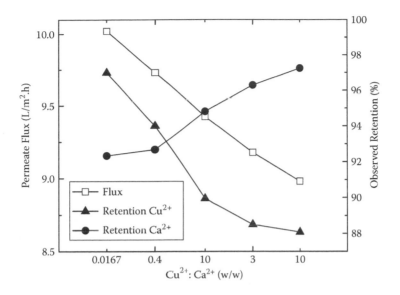

FIGURE 5.10
Effect of feed concentration on permeate flux and solute retention in the mixture of Ca^{2+} and Cu^{2+}; operating conditions TMP 345 kPa, SDS concentration 25 kg/m^3, cross-flow rate 60 LPH, and 5 kDa membrane.

A mixture of two heavy metal cations can be removed by MEUF using SDS. Figure 5.10 shows the retention of each of the ions and the permeate flux. It is observed that the permeate flux decreases as the total concentration of the salts increases. This indicates that the capability of both solutes to reduce gel concentration is working here in tandem, which decreases the flux values further (compared to a single solute system).[53] Interestingly, it may be noted that the retention of both the solutes is slightly less in the mixture (compared to single solute system). The slight decrease in the retention may be explained by the competitive solubilization of the two solutes. Solubilization of calcium is favored over that of copper. The retention of both solutes is unaffected by the TMP and cross-flow rate of the system. Quantification of the degree of fractional counterion binding by each of the ions on the micelle surface has been presented by Das et al.[54] using localized adsorption theory.

5.2.2 Anionic-Anionic Mixture

The presence of multiple toxic anions can be effectively removed by MEUF. Simultaneous removal of ferricyanide and nitrate using CPC has been studied by Baek et al.[55] They have reported that using a 10 kDa membrane (TMP 200 kPa), the removal of ferricyanide (nitrate) increased from 62% (<2%) to 99.9% (78%) when ferricyanide:nitrate:CPC was increased from 1:1:1 to 1:1:10 (both of the contaminants have an initial concentration of 1 mM). However, when the anions are individually subjected to removal by CPC, the removal efficiencies are >99% and 93% for ferricyanide and nitrate, respectively, using an S/M ratio of 1:10. In another study by Baek and Yang,[56] removal of nitrate and chromate using the CPC and cellulose acetate membrane having MWCO 3 to 10 kDa has been reported. The rejections of nitrate and chromate are 80 and 98%, respectively, at a nitrate:chromate:CPC ratio of 1:1:10 (TMP is 200 kPa). Simultaneous removal of a binary mixture of ferricyanide and chromate ions has been reported by Baek et al.[57] Their study reveals that using octadecylamine acetate (ODA) with a 10 kDa membrane and TMP at 200 TMP, the removal of ferricyanide (and chromate) increased from 62% (20%) to >99.9% (98%) when the ratio of ferricyanide:nitrate:ODA was increased from 1:1:1 to 1:1:6. This reveals the fact that as the valence of the anion increases, the binding capacity increases. Hence, in the case of a mixture of ferricyanide/nitrate/chromate, the ion having the highest valency (ferricyanide in this case) would have the maximum removal efficiency. However, at higher concentration, the ions would adhere to the micelle in the order of decreasing valence. Competitive inhibition would only occur when there is a limited proportion of the surfactant in the multiple-ion system.

Similar to the removal of ferricyanide and nitrate, simultaneous removal of two oxyanions, namely permanganate (MnO_4^-) and dichromate ($Cr_2O_7^{2-}$), from a waste stream by MEUF has been successfully carried out by Purkait et al.[58] The effect of feed surfactant concentration on the solute retention and permeate flux has been illustrated in Figure 5.11. It may be noted that

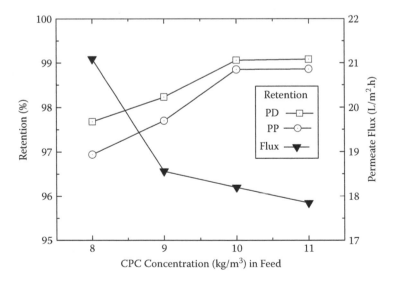

FIGURE 5.11
Effect of feed surfactant concentration on the retention and permeate flux; TMP 345 kPa, cross-flow rate 30 LPH. Solute concentrations: potassium dichromate (PD), 0.1 kg/m³; potassium permanganate (PP), 0.1 kg/m³.

flux decreases with the feed CPC concentration due to more concentration polarization over the membrane surface. As observed from Figure 5.11, the retention of both PD and PP increases with CPC concentration and remains almost constant beyond a surfactant concentration of 10 kg/m³. Therefore, 10 kg/m³ may be considered the optimum surfactant concentration for the separation of PD and PP in their mixture at a concentration of 0.1 kg/m³ each. The effect of pressure and cross-flow rate is not significant on the retention of the contaminants; however, retention of the surfactant CPC is increased with the cross-flow rate. Comparison of the retention results with batch and cross-flow reveals that at 10 kg/m³, the observed retention is almost the same. This indicates that solubilization of counterions on micelles is independent of mode of filtration.

5.2.3 Cationic-Anionic Mixture

Another potential application of MEUF is the simultaneous removal of a cation and an anion using the mixed micellar system. Variation of permeate flux and retention of a mixture of anion and cation have been presented in Figure 5.12.

In a mixed micellar system, the flux varies from 1 to 6 L/m².h. The intermicellar interaction of the oppositely charged micelles (negative for SDS and positive for CPC micelles) is quite strong in the gel type layer deposited over the membrane surface. The presence of both types of charged micelles

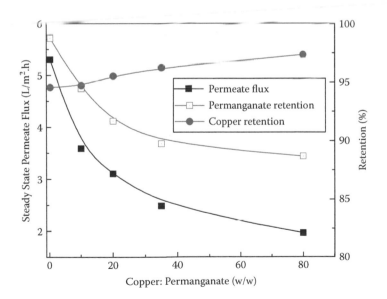

FIGURE 5.12 (See color insert.)
Removal of copper and permanganate using 25 kg/m³ SDS and 10 kg/m³ CPC; 5 kDa TFC membrane, TMP 414 kPa, and 30 LPH.

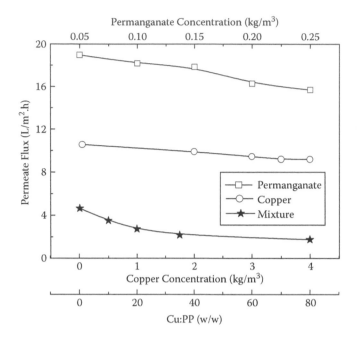

FIGURE 5.13
Variation of the permeate flux with single component and mixture using 25 kg/m³ SDS and 10 kg/m³ CPC; 5 kDa TFC membrane, TMP 414 kPa, and 30 LPH.

leads to the formation of a more compact gel layer with increased thickness. Hence, the flux of a mixed micellar system is much less than that of a single component system[59] (refer to Figure 5.13).

Variation of the cross-flow rate in the range of 30 to 60 LPH does not enhance the permeate flux, which signifies that the viscous resistance is very large compared to the flow rate driving force. Retention is unaffected by the change in TMP and cross-flow rate; it is governed solely by solubilization of counterions on the micellar surface. The observed retention values in the mixed micellar system are quantitatively the same as with the single component system; hence the solubilization capacity is independent in the presence of co-ions and operating conditions.[59]

Combined use of anionic and cationic surfactants is very useful in removal of different types of charged ions. In removal of a mixture of Cu^{2+}, PP, and PD, 25 and 10 kg/m³ of SDS and CPC have been used simultaneously.[60] The variation of permeate flux and retention with feed composition in a mixed micellar system is presented in Figures 5.14 and 5.15, respectively, for different feed compositions at a transmembrane pressure of 414 kPa, and cross-flow rate of 60 LPH. It is observed that the flux declines with an increase in the total solute concentration. For example, at a feed composition of 0.05 kg/m³ of each solute the flux is about 17.31 L/m².h, and it decreases to 11 L/m².h

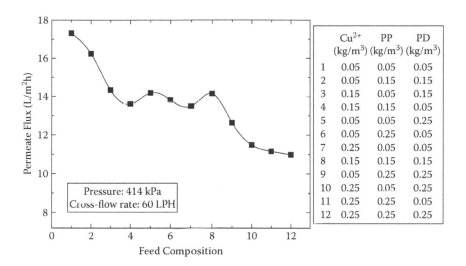

FIGURE 5.14
Variation of permeate flux with feed composition in mixed micellar system (5 kDa membrane).

at a total solute concentration of 0.75 kg/m³. This trend may be explained by the increase in resistance against the solvent flux due to micellar aggregates, which are deposited on the membrane surface. It may also be observed that the extents of retention of PD and PP are slightly different (more for PD and less for PP at the same operating condition). The slight difference in the

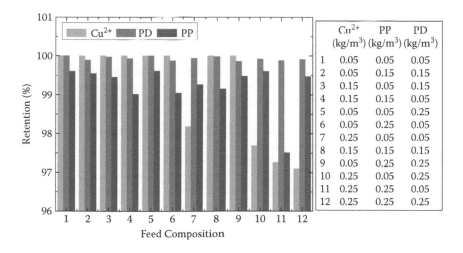

FIGURE 5.15 (See color insert.)
Effect of feed composition on retention of solutes; TMP 414 kPa, 5 kDa membrane, and cross-flow rate 60 LPH.

extent of the solubilization for PD and PP is due to their ionic characteristics. Attachment of a divalent oxyanion (PD) to the exterior of the micelles is more compared to that of the monovalent oxyanion (PP). Therefore, the permeate (containing unsolubilized solute) concentration of PD is lower and shows more observed retention.

References

1. Martell, A.E. 1981. Chemistry of carcinogenic metals. *Environ. Health Perspect.* 40: 207–226.
2. Antonious, G.F., Dennis, S.O., Unrine, J.M., and Synder, J.C. 2011. Heavy metals uptake in plant parts of sweet potato grown in soil fertilized with municipal sewage sludge. *Int. J. Geo.* 5: 14–20.
3. Zekri, O.C.A.Y., and Islam, R. 2005. Uptake of heavy metals by microorganisms: An experimental approach. *Energy Sources* 27: 87–100.
4. Li, C.-W., Liu, C.-K., and Yen, W.S. 2006. MEUF with mixed surfactants for removing Cu(II) ions. *Chemosphere* 63: 353–358.
5. Juang, R.-S., Lin, S.-H., and Peng, L.-C. 2010. Flux decline analysis in MEUF of synthetic waste solutions for metal removal. *Chem. Eng. J.* 161: 19–26.
6. Liu, C.-K., and Li, C.-W. 2005. Combined electrolysis and micellar enhanced ultrafiltration (MEUF) process for metal removal. *Sep. Purif. Technol.* 43: 25–31.
7. Huang, J.-H., Zeng, G.-M., Fang, Y.-Y., Qu, Y.-H, and Li, X. 2009. Removal of cadmium ions using micellar-enhanced ultrafiltration with mixed anionic-nonionic surfactants. *J. Membr. Sci.* 326: 303–309.
8. Ke, X., Guang-Ming, Z., Jin-Hui, H., et al. 2007. Removal of Cd²⁺ from synthetic wastewater using micellar-enhanced ultrafiltration with hollow fiber membrane. *Colloids Surf. A Physicochem. Eng. Aspects* 294: 140–146.
9. Fang, Y.-Y., Zenga, G.-M., Huang, J.-H., et al. 2008. Micellar-enhanced ultrafiltration of cadmium ions with anionic–nonionic surfactants. *J. Membr. Sci.* 320: 314–319.
10. Landaburu-Aguirre, J., Pongrácz, E., Perämäki, P., and Keiski, R.L. 2010. MEUF for the removal of Cd²⁺ and Zn²⁺: Use of response surface methodology to improve understanding of process performance and optimization. *J. Hazard. Mater.* 180: 524–534.
11. Ghosh, G., and Bhattacharya, P.K. 2006. Hexavalent chromium ion removal through MEUF. *Chem. Eng. J.* 119: 45–53.
12. Gzara, L., and Dhahbi, M. 2001. Removal of chromate anions by micellar-enhanced ultrafiltration using cationic surfactants. *Desalination* 137: 241–250.
13. Aoudia, M., Allal, N., Djennet, A., and Toumi, L. 2003. Dynamic MEUF: Use of anionic (SDS)–non-ionic (NPE) system to remove Cr³⁺ at low surfactant concentration. *J. Membr. Sci.* 217: 181–192.
14. Chhatre, A.J., and Marathe, K.V. 2006. Dynamic analysis and optimization of surfactant dosage in micellar enhanced ultrafiltration of nickel from aqueous streams. *Sep. Sci. Technol.* 41: 2755–2770.
15. Yurlova, L., Kryvoruchko, A., and Kornilovich, B. 2002. Removal of Ni(II) ions from wastewater by MEUF. *Desalination* 144: 255–260.

16. Danis, U., and Aydiner, C. 2009. Investigation of process performance and fouling mechanisms in micellar-enhanced ultrafiltration of nickel-contaminated waters. *J. Hazard. Mater.* 162: 577–587.
17. Danis, U., and Aydiner, C. 2009. Investigation of process performance and fouling mechanisms in micellar-enhanced ultrafiltration of nickel-contaminated waters. *J. Hazard. Mater.* 162: 577–587.
18. Mondal, S., and De, S. 2010. A fouling model for steady state crossflow membrane filtration considering sequential intermediate pore blocking and cake formation. *Sep. Purif. Technol.* 75: 222–228.
19. Rahmanian, B., Pakizeh, M., and Maskooki, A. 2010. Micellar-enhanced ultrafiltration of zinc in synthetic wastewater using spiral-wound membrane. *J. Hazard. Mater.* 184: 261–267.
20. Zhang, Z., Zeng, G.-M., Huang, J.-H., et al. 2009. Removal of zinc ions from aqueous solution using MEUF at low surfactant concentrations. *Water SA* 33: 129–136.
21. Huang, J.-H., Zeng, G.-M., Qu, Y.-H., and Zhang, Z. 2007. Adsorption characteristics of zinc ions on sodium dodecyl sulfate in process of MEUF. *Trans. Nonferrous Met. Soc. China* 17: 1112–1117.
22. Akita, S., Yang, L., and Takeuchi, H. 1997. Micellar-enhanced ultrafiltration of gold(III) with nonionic surfactant. *J. Membr. Sci.* 133: 189–194.
23. Warshawsky, A. 1984. Hydrometallurgical processes for the separation of platinum group metals (PGM) in chloride media. In *Ion exchange technology*, 604–610. Chichester: Ellis Horwood Ltd.
24. Yenphan, P., Chanachai, A., and Jiraratananon, R. 2010. Experimental study on micellar-enhanced ultrafiltration (MEUF) of aqueous solution and wastewater containing lead ion with mixed surfactants. *Desalination* 253: 30–37.
25. Ghezzi, L., Robinson, B.H., Secco, F., Tiné, M.R., and Venturini, M. 2008. Removal and recovery of palladium(II) ions from water using micellar-enhanced ultrafiltration with a cationic surfactant. *Colloids Surf. A Physicochem. Eng. Aspects* 329: 12–17.
26. Tillier-Dorion, F., Charbit, G., and Gaboriaud, R. 1987. Comparison between effects of micellized alkyl sulfates and of dissociated electrolytes upon activity coefficients. *J. Colloid Interf. Sci.* 116: 588.
27. Gaboriaud, R., Charbit, G., and Dorion, F. 1984. Acidic lauryl sulfate synthesis: Degree of proton association measurement in alkyl sulfate micelles. *J. Colloid Interf. Sci.* 98: 583–584.
28. Baes, C.F., and Mesmer, R.E. 1976. *The hydrolysis of cations.* New York: John Wiley & Sons.
29. Gwicana, S., Vorstera, N., and Jacobs, E. 2006. The use of a cationic surfactant for micellar-enhanced ultrafiltration of platinum group metal anions. *Desalination* 199: 504–506.
30. Rajec, P., and Paulenova, A. 1994. Micellar enhanced microfiltration of strontium. *J. Radioanal. Nucl. Chem.* 183: 109–113.
31. Shaw, D.J. 1980. *Introduction to colloid and surface chemistry.* 3rd ed. London: Butterworths.
32. Iqbal, J., Kim, H.-J., Yang, J.-S., Baek, K., and Yang, J.-W. 2007. Removal of arsenic from groundwater by micellar enhanced ultrafiltration (MEUF). *Chemosphere* 66: 970–976.
33. Beolchini, F., Pagnanelli, F., Michelis, I.D., and Veglio, F. 2006. Micellar enhanced ultrafiltration for arsenic(V) removal: Effect of main operating conditions and dynamic modelling. *Environ. Sci. Technol.* 40: 2746–2752.

34. Beolchini, F., Pagnanelli, F., and Veglio, F. 2007. Treatment of concentrated arsenic(V) solutions by micellar enhanced ultrafiltration with high molecular weight cut-off membrane. *J. Hazard. Mater.* 148: 116–121.
35. Gecol, H., Ergican, E., and Fuchs, A. 2004. Molecular level separation of arsenic(V) from water using cationic surfactant micelles and ultrafiltration membrane. *J. Membr. Sci.* 241: 105–119.
36. Ergican, E., Gecol, H., and Fuchs, A. 2005. The effect of co-occurring solutes on the removal of As(V) from water using cationic surfactant micelles and UF membranes. *Desalination* 181: 9–26.
37. Maiti, A., Dasgupta, S., Basu, J.K., and De, S. 2007. Adsorption of arsenite using natural laterite as adsorbent. *Sep. Purif. Technol.* 55: 350–359.
38. Maiti, A., Sharma, H., Basu, J.K., and De, S. 2009. Modeling of arsenic adsorption kinetics of synthetic and contaminated groundwater on natural laterite. *J. Hazard. Mater.* 172: 928–934.
39. Roach, J.D., and Zapien, J.H. 2009. Inorganic ligand-modified, colloid-enhanced ultrafiltration: A novel method for removing uranium from aqueous solution. *Water Res.* 43: 4751–4759.
40. Pramauro, E., and Pehzzetti, E. 1988. Micelles: A new dimension in analytical chemistry. *Trends Anal. Chem.* 7: 260–265.
41. Pramauro, E., Prevot, A.B., Pehzzetti, E., Marchelli, R., Dossena, A., and Biancardi, A. 1992. Quantitative removal of uranyl ions from aqueous solutions using micellar-enhanced ultrafiltration. *Anal. Chimica Acta* 264: 303–310.
42. Reiller, P., Lemordant, D., Moulin, C., and Beaucaire, C. 1994. Dual use of micellar enhanced ultrafiltration and time-resolved laser-induced spectrofluorimetry for the study of uranyl exchange at the surface of alkylsulfate micelles. *J. Colloid Interf. Sci.* 163: 81–86.
43. Gadelle, F., Wan, J., and Tokunaga, T.K. 2001. Removal of uranium(VI) from contaminated sediments by surfactants. *J. Environ. Qual.* 30: 470–478.
44. Woodard and Curran Inc. 2000. *Industrial waste treatment handbook*, Oxford: Butterworth-Heinemann.
45. Wang, L.K., Hung, Y.-T., and Lo, H.H., eds. 2007. *Hazardous industrial waste treatment*. Boca Raton, FL: Taylor & Francis.
46. Sheng, P.X., Ting, Y.-P., and Chen, J.P. 2007. Biosorption of heavy metal ions (Pb, Cu, and Cd) from aqueous solutions by the marine alga *Sargassum* sp. in single- and multiple-metal systems. *Ind. Eng. Chem. Res.* 46: 2438–2444.
47. Kadirvelu, K., Thamaraiselvi, K., and Namasivayam, C. 2001. Removal of heavy metals from industrial wastewaters by adsorption onto activated carbon prepared from an agricultural solid waste. *Bio. Res. Technol.* 76: 63–65.
48. Channarong, B., Lee, S.H., Bade, R., and Shipin, O.V. 2010. Simultaneous removal of nickel and zinc from aqueous solution by micellar-enhanced ultrafiltration and activated carbon fiber hybrid process. *Desalination* 262: 221–227.
49. Karate, V.D., and Marathe, K.V. 2008. Simultaneous removal of nickel and cobalt from aqueous stream by cross flow micellar enhanced ultrafiltration. *J. Hazard. Mater.* 157: 464–471.
50. Das, C., Dasgupta, S., and De, S. 2008. Prediction of permeate flux and counter-ion binding during cross-flow micellar-enhanced ultrafiltration, *Colloids Surf. A Physicochem. Eng. Aspects* 318: 125–133.

51. Juang, R.S., Xu, Y.Y., and Chen, C.L. 2003. Separation and removal of metal ions from dilute solutions using micellar-enhanced ultrafiltration. *J. Membr. Sci.* 218: 257–267.
52. Scamehorn, J.F., Christian, S.D., Sayed, D.A., Uchiyama, H., and Younis, S.S. 1994. Removal of divalent metal cations and their mixtures from aqueous streams using micellar-enhanced ultrafiltration. *Sep. Sci. Technol.* 29: 809–830.
53. Urbański, R., Goralska, E., Bart, H.J., and Szymanowski, J. 2002. Ultrafiltration of surfactant solutions. *J. Colloid Interf. Sci.* 253: 419–426.
54. Das, C., Dasgupta, S., and De, S. 2008. Prediction of permeate flux and counter-ion binding during cross-flow micellar-enhanced ultrafiltration. *Colloids Surf. A Physicochem. Eng. Aspects* 318: 125–133.
55. Baek, K., Lee, H.-H., and Yang, J.-W. 2003. Micellar-enhanced ultrafiltration for simultaneous removal of ferricyanide and nitrate. *Desalination* 158: 157–166.
56. Baek, K., and Yang, J.-W. 2004. Micellar-enhanced ultrafiltration of chromate and nitrate: Binding competition between chromate and nitrate. *Desalination* 167: 111–118.
57. Baek, K., Kim, B.-K., Cho, H.-J., and Yan, J.-W. 2003. Removal characteristics of anionic metals by micellar-enhanced ultrafiltration. *J. Hazard. Mater.* B99: 303–311.
58. Purkait, M., Dasgupta, S., and De, S. 2005. Simultaneous separation of two oxyanions from their mixture using micellar enhanced ultrafiltration. *Sep. Sci. Technol.* 40: 1439–1460.
59. Das, C., Maity, P., Dasgupta, S., and De, S. 2008. Separation of cation–anion mixture using micellar-enhanced ultrafiltration in a mixed micellar system. *Chem. Eng. J.* 144: 35–41.
60. Das, C. 2009. Treatment of tannery effluent and removal of pollutants using micellar enhanced ultrafiltration. PhD diss., Indian Institute of Technology Kharagpur.

6

Removal of Organic Pollutants

The Clean Water Act (CWA) of the U.S. government listed 126 priority pollutants in wastewater discharges, which have since been standardized by the USEPA as major chemical pollutants that need to be regulated and monitored. Interestingly, a large proportion of these priority pollutants includes organic components, most of which are aromatic in nature. It simply affirms the fact that the organic contaminants contribute severe environmental pollution. Typical physiological problems associated with organic components have high carcinogenic effects, liver problems, nervous effects, and reproductive issues. Table 6.1 presents some of these toxic chemicals, their permissible limits in drinking water, and potential health hazards.

6.1 Removal of Single Component System

6.1.1 Removal of Dyes

Among the organic pollutants, dyes constitute a significant proportion of the waste in the effluent stream. The chemical structure of dyes constitutes an aromatic ring in general. The chemical structures of some of the typical dyes present in waste effluent are represented in Figure 6.1.

The current process of removing synthetic dyes includes adsorption on various sorbents,[1] chemical decomposition by oxidation,[2,3] photodegradation,[4] microbiological decoloration, activated sludge[5], coagulation/flocculation[6] pure cultures, and microbe consortiums. Among the membrane-based processes, nanofiltration (NF)[7] and reverse osmosis (RO) are the best available commercial techniques. The major disadvantage of these processes is the decline in permeate flux to achieve high selectivity. Similar to removal of metals, surfactant-based separation methods are also effective in removal of dyes and organic contaminants. Removal of dye by solubilization using cetylpyridinium chloride (CPC) has been first reported by Purkait et al.[8] In the separation of eosin dye, an organic polyamide membrane of molecular weight cutoff (MWCO) 1 kDa has been used in unstirred batch mode. It is worth noting that retention of the dye (eosin) increases from 10% to 73.4% using surfactant micelles in the concentration range of 1 to 20 kg/m^3 at 276 kPa (refer to Figure 6.2). The cross-flow experiments reveal that the

TABLE 6.1

Primary Standards of Drinking Water Regulation

Organic Contaminant	Permissible Limit in Drinking Water (ppm)	Potential Health Hazards
Alachlor	0.002	Eye, liver, kidney, or spleen problems; anemia; increased risk of cancer
Benzene	0.005	Anemia; decrease in blood platelets; increased risk of cancer
Benzo(a)pyrene (PAHs)	0.0002	Reproductive difficulties; increased risk of cancer
Chlorobenzene	0.1	Liver or kidney problems
Toluene	1	Nervous system, kidney, or liver problems
Vinyl chloride	0.002	Increased risk of cancer
Styrene	0.1	Liver, kidney, or circulatory system problems
Ethylbenzene	0.7	Liver or kidney problems
Di(2-ethylhexyl) phthalate	0.006	Reproductive difficulties; liver problems; increased risk of cancer
Chlorophenol	0.001	Liver or kidney problems; increased cancer risk
Xylene	10	Nervous system damage
o-Dichlorobenzene	0.6	Liver, kidney, or circulatory system problems
p-Dichlorobenzene	0.075	Anemia; liver, kidney, or spleen damage; changes in blood
Hexachlorobenzene	0.001	Liver or kidney problems; reproductive difficulties; increased risk of cancer

Source: USEPA. 816-F-09-0004, May 2009.

observed retention of eosin (maximum about 74%) is almost independent of the operating pressure, whereas the permeate flux increases significantly with pressure. The permeate flux increases marginally (about 4.5%) at higher cross-flow rates.[9] In another study by Purkait et al.,[10] quantification of the various pore blocking and cake formation resistances has been carried out. The analysis reveals that at low pressure and during the initial few moments of the ultrafiltration, the process is mainly governed by the pore blocking phenomena, followed by cake formation.

Micellar-enhanced ultrafiltration (MEUF) of reactive dyes from aqueous solution, like C.I. Reactive Black 5 (RB5) and C.I. Reactive Orange 16 (RO16), was analyzed by Ahmad et al.[11] A polymeric membrane of 10 kDa cutoff has been used in dead-end ultrafiltration at 300 kPa. The result shows that the highest dye rejections for RB_5 and RO_{16} dyes are 99.7 and 99.6% for 1.000 g/L CPC and 0.050 g/L dye concentrations, respectively, as shown in Figure 6.3. It is worthwhile to mention that Koyuncu has reported a

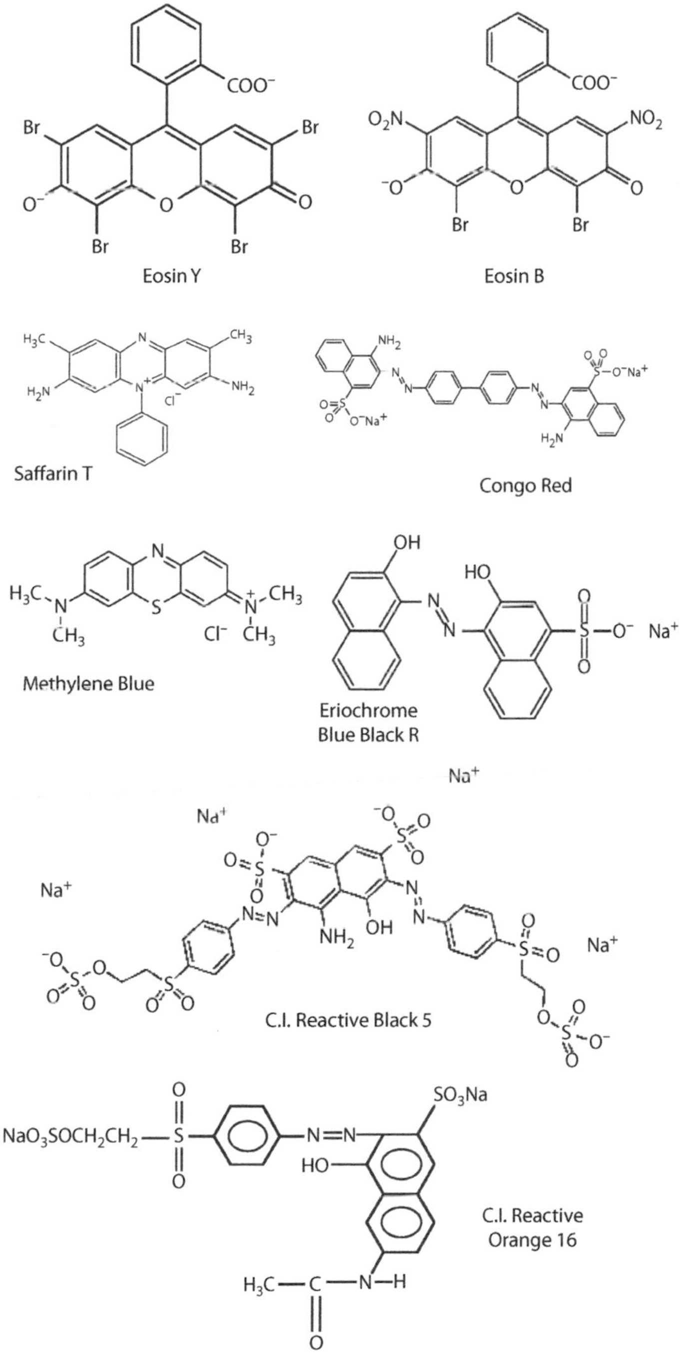

FIGURE 6.1
Chemical structure of some of the typical dyes present in wastewater.

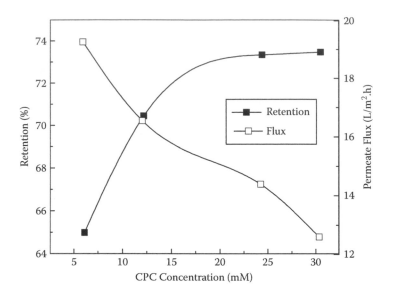

FIGURE 6.2
Effect of CPC concentration on the retention and permeate flux of MEUF of eosin dye; feed concentration 10 ppm, TMP 276 kPa, and 1 kDa membrane.

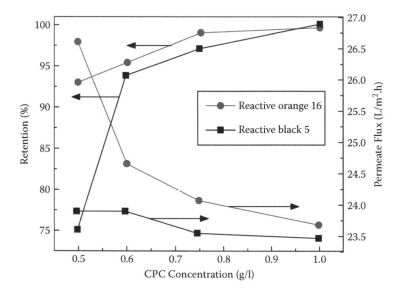

FIGURE 6.3 (See color insert.)
Variation of rejection and permeate flux at different CPC concentrations for MEUF of reactive dyes using 10 kDa membrane, TMP 300 kPa, and feed concentration 50 ppm.

maximum removal efficiency of RB_5 and RO_{16} of 99.5% using an NF membrane having MWCO (150 to 300 Da) operating at 2,400 kPa.[12] This clearly indicates the potential interest of MEUF in the present context.

Separation of methylene blue from aqueous solution has been studied by Zaghbani et al.[13] using a sodium dodecyl sulfate (SDS). The experiments reveal that using a cellulose membrane of 10 kDa cutoff, the maximum retention of methylene blue can be achieved within 95 to 99% at SDS concentration of 10 mM and above, using the cross-flow mode. However, increasing the SDS concentration beyond 10 mM decreases the permeate flux significantly (refer to Figure 6.4). The effect of pH is not significant in the retention characteristics. However, the permeate flux increases from 42 to 65 $L/m^2.h$ when pH is increased from 2 to 11. Using a mixed micellar system of SDS and oxyethylated coconut fatty acid methyl esters (OMC-10) in the ratio of 4:1, respectively, a retention of 93 to 94% can be achieved, which also reduces the surfactant concentration in the permeate stream by 75% compared to UF using only SDS. It should be noted here that when the nonionic surfactant (OMC) is used alone, the retention is only 34% due to the absence of ion-pair formation with the cationic dye.[14] The membranes used for the experimental analysis are composed of cellulose, polyether sulfone (PES), and polyvinylidene fluoride (PVDF) having MWCO 15 kDa to 30 kDa. In another study by Pozniak et al.,[15] using two times the critical micellar concentration (CMC) of SDS gives satisfactory levels of removal of methylene blue using modified PES membranes.

Eriochrome Blue Black R (EBBR) can also be removed by micellar-enhanced ultrafiltration using an alkyltrimethylammonium bromide of different chain lengths. It has been shown by Zaghbani et al.[13] that as the chain length increases, the retention increases significantly. The retention values of EBBR (1 mM), when ultrafiltered in cross-flow mode, are 34.6, 51.1, 99.6, and 99.9% in the presence of $C_{12}TAB$, $C_{14}TAB$, $C_{16}TAB$, and $C_{18}TAB$, respectively, at 10 times the respective CMC. The permeate flux increases significantly on increasing the TMP, without any change in the retention. However, there is one interesting point to observe: in the presence of electrolytes (Na_2SO_4, NaCl, Na_2HPO_4), the retention is hardly affected, which is unlike the MEUF of inorganic contaminants. This is probably due to solubilization of EBBR in a region between the core and the surface of the micelle. Large polar molecules, such as polar dyestuffs, are believed to be solubilized, in aqueous medium, mainly between the individual molecules of the surfactant in the palisade layer, with the polar groups of the solubilizate oriented toward the polar groups of the surfactants and the nonpolar moieties oriented toward the interior of the micelle.[16]

Removal of Saffarin T from wastewater streams has been carried out by Zaghbani et al.,[17] using a cellulose membrane of 10 kDa MWCO in cross-flow mode. The rejection efficiency is 99% in the range of dye concentration of 0.02 to 5 mM using an SDS concentration varying from 10 to 100 mM (see Figure 6.4). The permeate flux increases with TMP and decreases with the concentration of added electrolyte. For example, the permeate flux decreases

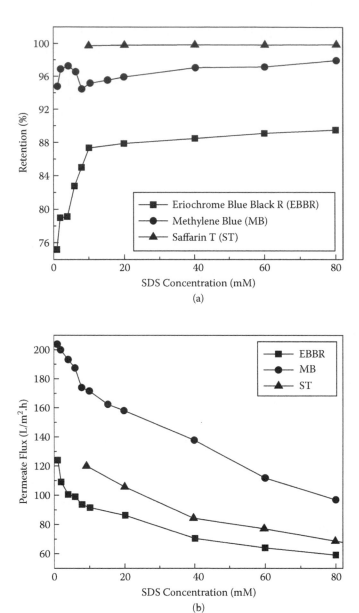

FIGURE 6.4
Effect of SDS concentration on the retention of dyes and permeate flux; operating conditions: TMP 1.4 bar, MWCO 10 kDa, feed concentration [EBBR] 1 mM, [MB] 1 mM, and [ST] 0.5 mM.

from 114 to 64 $L/m^2.h$ when the NaCl concentration increases from 0.01 to 0.5 mM (SDS concentration 10 mM) and TMP is 1.4 bar. Unlike the retention of inorganic pollutants, the organic contaminants are quite insensitive to change in pH.

6.1.2 Removal of Phenol

Another important pollutant in the wastewater stream is phenol, which is present in industrial effluent of petroleum refineries, plastic manufacturing plants, pharmaceutical industries, coal carbonization and tar distillation units, wood charcoal production units, coke ovens, phenol-formaldehyde plants, and bis-phenol A and other synthetic resin manufacturing units. It is present abundantly in the antiseptic and germicide solution used for household purposes. The conventional techniques include biodegradation by immobilization of specific cultured microorganisms, adsorption by activated carbon processes,[18] and semifluidized bed bioreactor.[19] However, all these processes are cost-intensive and time-consuming compared to micellar-enhanced ultrafiltration.

Removal of phenol (P) and its other derivatives (para-nitrophenol (PNP), meta-nitrophenol (MNP), beta-napthol (BN), catechol (CC), and ortho-chlorophenol (OCP)) has been carried out by Purkait et al.[20] using CPC. At the end of the batch experiment, with a surfactant-to-solute ratio of 110, the observed retentions are 66, 78, 94, 96, 96, and 98% for P, CC, OCP, PNP, MNP, and BN, respectively, when operated at 345 kPa. Comparison of cross-flow with batch MEUF is presented in Figure 6.5.

MEUF of phenol in unstirred mode has been carried out by Syamal et al.[21] using a 1 kDa membrane and CPC as surfactant. It has been reported that an optimum S/M of 30 exists typically for MEUF at varying TMP. The rejection efficiency of phenol decreases with operating pressure, varying from 84% to 72% as TMP is increased from 207 kPa to 483 kPa at an S/M ratio of 30. Zeng et al.[22] have shown ultrafiltration studies for removal of phenol using CPC in cross-flow mode. Their work shows that the removal efficiency increases significantly on increasing the CPC concentration up to 30 mM operating at a TMP of 150 kPa and feed phenol concentration of 1 mM. However, at low CPC concentration, the surfactant rejection is lower. They have reported that using two different MWCO PES membranes of 6 and 10 kDa, the rejection is almost equal at a CPC concentration beyond 30 mM, since the permeate flux is much higher, considering a 10 kDa membrane. Hence, it is recommended to use a high flux membrane at a higher CPC concentration. Also, they have observed that on increasing temperature, rejection decreases considerably. In another study by Sabate et al.,[23] MEUF of phenol using ceramic and PES membranes has been studied. It explains that the maximum rejection of phenol and the surfactant are independent of the type of membrane material used. However, depending on the membrane material, the optimized set of parameters for MEUF is affected. The development of the resistances over

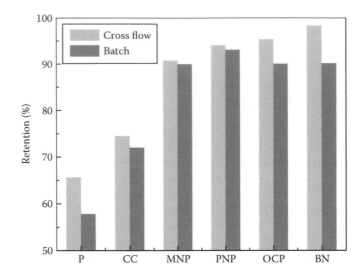

FIGURE 6.5 (See color insert.)
Relative retention of different phenols using 10 kg/m³ CPC and 0.24 kg/m³ solute concentration; TMP 345 kPa, 1 kDa membrane and concentration, TMP 345 kPa, 1 kDa membrane, and 30 LPH cross-flow rate.

the membrane surface grows significantly with time of ultrafiltration. But the relation of permeate flux with resistance is independent of the concentration of the surfactant/phenol and chemical nature of the pollutant, since, in general, the steady-state resistance ultimately converges to the same magnitude for different surfactant/phenol concentrations.[24] The relative orders of effectiveness of various surfactants, i.e., cetyltrimethylammonium bromide (CTAB), SDS, alkylpolyglucoside (APG), and oxyethylated methyl dodecanoate with an average degree of oxyethylation 5 (OMD-5) and 9 (OMD-9) for removal of phenol and *m*-nitrophenol, have been analyzed by Adamczak et al.[25] The effects of different surfactants on flux and rejection efficiency are summarized in Table 6.2.

TABLE 6.2
Relative Effectiveness of the Various Surfactants When Used at High Surfactant Concentrations

Substrate	Flux	Rejection Efficiency
Phenol	APG > OMD-5 ≥ SDS > CTAB > OMD-9	CTAB > SDS >> APG > OMD-5 > OMD-9
m-Nitrophenol	CTAB > APG > OMD-5 > OMD-9 = SDS	CTAB >> OMD-5 > APG = SDS = OMD-9

Bielska et al.,[26] in the removal of nitrophenol, have used CTAB, oxyethylated coconut fatty acid methyl esters of an average oxyethylation degree equal to 10 (OMC-10), and oxyethylated nonyl phenol of an average oxyethylation degree equal to 9 (ONP-9) in polymeric membranes (MWCO 15 to 30 kDa). Their result shows that the average rejection of nitrophenol is above 90% using a PES membrane with CTAB and mixed micellar systems at five times CMC. However, the average permeate flux of a cellulosic membrane is better than that of the other polymeric membranes. However, in one instance, considering both the average permeate flux and the rejection using the PES membrane and a mixed micellar system of CTAB + ONP-9, a maximum rejection and permeate flux are produced. Separation of phenol and nitrophenol can be explained by the generalized linear solvation theory proposed by Abraham et al.[27,28]

n-Alcohols can be removed effectively by MEUF. The rejection efficiency improves by increasing the chain length of the alkyl alcohol and S/M ratio[29] as illustrated in Figure 6.6. A hydrophobic chain increases the hydrophobicity of the alcohol molecules, resulting in more solubilization by the micelle.[30] At a constant S/M ratio of 10, the permeate concentration decreases by addition of a methylene group in the linear alkyl chain of the alcohol[29] (refer to Figure 6.6). For a particular *n*-alkyl alcohol, the permeate concentration varies almost linearly with increasing S/M ratio.[31]

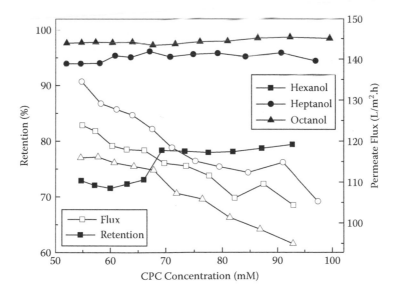

FIGURE 6.6
Effect of CPC concentration on MEUF of *n*-alcohols; TMP 414 kPa, 10 kDa membrane, initial S/M ratio of 10.

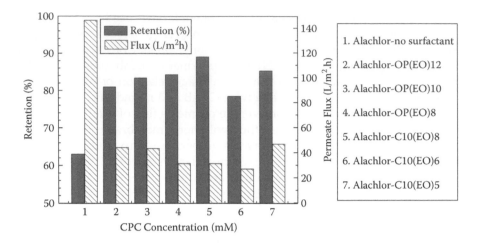

FIGURE 6.7
Micellar-enhanced ultrafiltration of alachlor using different nonionic surfactants at 2 CMC surfactant concentration, 70 ppm solute concentration, TMP 1 bar, and 10 kDa PES membrane.[38]

6.1.3 Removal of Alachlor

Alachlor is one of the high-priority toxic chemicals present in pesticide, and contaminates groundwater by surface runoff, soil degradation, and leaching. According to the USEPA, alachlor has a high risk factor of carcinogenic effect in humans. Previous techniques of removal of alachlor involve usage of activated carbon adsorption,[32] catalyzed ozonization,[33] treatment with zerovalent iron,[34] etc. Micellar-enhanced ultrafiltration also offers a potential alternate solution. The effect of nonionic surfactant, namely, octa phenol ethoxylates [OP(EO)$_n$] and decyl alcohol ehtoxylates [C10(EO)$_n$], having different numbers of ethoxy groups, is presented in Figure 6.7. It has already been reported in the literature that alachlor is solubilized in nonionic surfactant micelles.[35] The properties of these surfactants are presented in Table 6.3. It can be observed

TABLE 6.3
Properties of Nonionic Surfactants Used for the Removal of Alachlor

Name	Molecular Weight (g/mol)	CMC (mM)
Octa Phenol Ethoxylates		
OP(EO)8	558	0.265
OP(EO)10	646	0.280
OP(EO)12	756	0.370
Decyl Alcohol Ethoxylates		
C10(EO)5	380	0.8
C10(EO)6	424	0.9
C10(EO)8	512	1.0

from the figure that, using OP(EO)8, maximum retention (41% increase) of alachlor is achieved. It must be noticed that the surfactant concentration used was only two times CMC, which is responsible for lower retention values, compared to other organic contaminants. Even though the molecular weight of alachlor is 269.8 g/mol, the relatively high rejection of alachlor without the use of surfactant in 10 kDa membrane suggests that adsorption occurs on a polymeric membrane surface. It has been shown in the literature that organic molecules adsorb on the surface of a membrane even at low concentration.[36,37]

6.1.4 Removal of Naphthenic Acid

Naphthenic acids are the most significant environmental contaminants resulting from petroleum extraction from oil sand deposits.[39] Naphthenic acid (NA) is the name for an unspecific mixture of several cyclopentyl and cyclohexyl carboxylic acids with molecular weights ranging from 120 to well over 700. The presence of NA in local water bodies and their potential effects on water quality and fish reproduction and tainting have brought significant attention to their persistence in the environment and to their aquatic toxicity at the levels found in tailings lakes.[40] NA causes tailings to be acutely toxic to aquatic organisms and mammals.[41] Naphthenic acid is carcinogenic in nature.[42] In addition to being acutely toxic, NA associated with oil sand tailings does not easily break down in the natural environment.[43] The retention of NA has a negative effect depending on the solution pH. Retention of NA decreases from 82% to 68% as the solution pH increases from 3 to 9. The pK value of NA is between 5 and 6, depending on its source. Thus, at lower pH (less than 5), NA becomes more cationic in nature and the whole molecule quickly aligns in the micelles with the hydrophobic part inside and polar part outside. This enables more efficient solubilization of NA in the SDS micelles at lower pH, resulting in higher retention of NA.

Optimum surfactant concentration was selected by performing experiments at pressure drop 414 kPa, cross-flow rate 60 L/h, 500 mg/l of feed NA, and pH 3.0 using different concentrations of surfactant varying from 2 CMC to 10 CMC. The results at the steady state are presented in Figure 6.8. It is observed from this figure that the retention of NA increases from 68 to 97% as the surfactant concentration increases from 2 CMC to 10 CMC. At 8 CMC, retention was about 96%. As more surfactants are present, the micelle concentration also increases, thereby, solubilizing more NA, and finally retention of NA becomes high. However, at higher surfactant concentration, retention of NA becomes almost invariant due to saturation of solubilization of NA in micelles. Surfactant retention also increases from 78% to 94% with surfactant concentration in feed. Surfactant concentration in the permeate is slightly less than CMC. Thus, as the feed concentration increases, observed retention of the surfactant also increases. On the other hand, the steady-state flux decreases with surfactant concentration. At higher concentration, the concentration of micelles becomes high. Thus, more micelles deposit on the

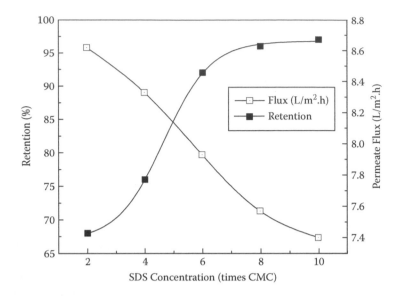

FIGURE 6.8
Effect of SDS concentration on the retention and permeate flux of MEUF of naphthenic acid. Operating conditions: pH 3.0, TMP 414 kPa, cross-flow rate 60 LPH, solute concentration 500 ppm, and 10 kDa membrane.

membrane surface, forming a thicker layer of micelles, thereby increasing the resistances against the solvent flux, resulting in lower permeate flux. Therefore, at higher surfactant concentration, an increase in retention of NA is associated with a decrease in permeate flux. Flux enhancement can be achieved by changing the TMP or application external field; however, solute retention is entirely governed by the solubilization of NA in surfactant micelles and solution chemistry (pH and electrolyte concentration).[44]

6.2 Removal of Multicomponent System (Exclusively Organic)

Similar to the removal of multiple inorganic pollutants, the micellar-enhanced ultrafiltration is quite efficient in removal of organic pollutants. Removal of low molecular weight molecules from wastewater has been first investigated by Dunn et al.[45] They have reported that six organic solutes (in low concentration) can be simultaneously removed with high removal efficiency using MEUF. Using cationic surfactant CPCs of 50 mM and 0.5 mM of the organic contaminants, the aqueous solution is ultrafiltered at 414 kPa and 1 kDa MWCO membrane. Since the micelles are completely rejected beyond a certain cutoff of the membrane there is development of the significant

TABLE 6.4

Rejection of Different Solutes Using CPC[31]

Solutes	Rejection (%)
Benzene	91.02
Cyclohexane	99.09
Phenol	94.02
p-Cresol	98.02
n-Pentanol	79.15
Chlorobenzene	97.63

concentration polarization over the surface of the membrane, which dictates the flux behavior of the system, thereby requiring an optimum set of the parameters for MEUF application.[46] Rejection of various organics is presented in Table 6.4 for an organic-CPC system.

Dissociation of organic pollutants is very sensitive to the ionic environment of the solution. Phenols are weak acids that can dissociate at high pH; however, amines are protonated in a low pH environment. Figure 6.9 presents the ionic nature of the phenolic and amine species at different acidic or alkaline environments. The use of ionic surfactants, both cationic and anionic, is effective when the solutes exist in ionic form in the solution. At lower pH, the amines are completely ionic and are very suitable for selective removal using ionic surfactant. In the neutral pH zone (pH 7), species 3, 4,

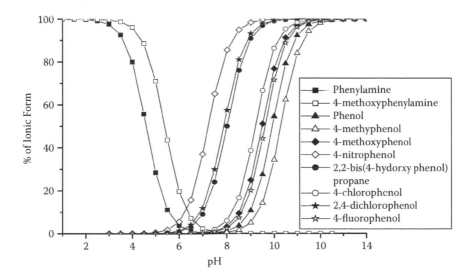

FIGURE 6.9

pH effect on the concentration of pollutant ionic species in pure aqueous solutions.[48] (Reproduced by permission of the Royal Society of Chemistry.)

5, and 9 are present in nonionic forms. The nonionic species are dominant for a pH range of 7 to 9, but the concentration of phenolic compounds can be estimated at the level of 10 to 20% in pure water and around 40% for 6. Species 1 is in nonionic form, but a small amount of protonation forms is exhibited by 2. In the alkaline range (pH 10), species 1 and 2 are in neutral forms. However, only 7 to 9 are dissociated. Other phenols, 3 to 5, exist in both nonionic and anionic forms.

Actually, the situation is more complicated in micellar solutions containing CTAB and SDS. Charged micelles of the cationic surfactant accumulate hydroxyl groups in the Stern layer. Thus, the local pH at the micelle interface is higher, compared with the one measured in solution. As a result, the apparent binding efficiencies increase and enhanced dissociation of weak acids (phenols) and protonated amines is observed. The opposite effect is observed in micellar solutions of anionic surfactants.[47]

All these indicate that, depending on the pH and the pollutant, one can expect the binding by the micelles of nonionic or ionic species or the binding of both forms.[48]

Simultaneous removal of phenol and aniline using a cationic surfactant CPC has been carried out by Jadhav et al.[49] Figure 6.10 shows that the percentage removal of phenol is more compared to that of aniline for all CPC concentrations. For example, at a CPC concentration of 85 mM (phenol and aniline feed concentrations are 3 mM each), the removal efficiencies are 85

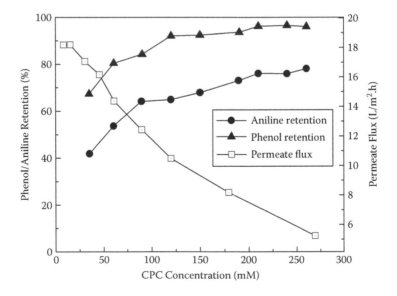

FIGURE 6.10
Variation of CPC concentration on the removal of aniline and phenol; operating conditions: TMP 450 kPa, 1 kDa membrane, feed aniline and phenol concentration 3 mM.

and 64% for phenol and aniline, respectively, using a 1 kDa membrane at 450 kPa TMP. It must be emphasized that since phenol is an ionic solute, it is solubilized on the surface of the micelle, thereby limiting the scope (increasing resistance) of the intramicellar diffusion of the nonionic solute (aniline in this case) into the core of the micelle. Hence, degree of solubilization is more in the case of phenol than aniline, resulting in a higher removal rate of phenol in the binary mixtures. With the increase in the surfactant concentration the flux decreases due to the development of a concentration polarization layer over the membrane surface by the solubilized micelles.

6.3 Organic-Inorganic Mixture

In the case of most industrial effluents, both organic and metal ions are present in the effluent stream. But, using a single surfactant, either a combination of nonionic (or nucleophillic) organic/anion or a nonionic (or electrophillic) organic/cation system can be handled. Counterintuitive combinations of electrophillic or nucleophillic organics along with an anion or cation require application of a mixed micellar system. Das et al.[50] have analyzed the flux and retention characteristics of MEUF of copper and β-napthol (BN) solution using SDS (25 kg/m^3), in a cross-flow system (cross-flow velocity 1.67 × 10^{-5} m^3/s) at TMP 276 kPa. An organic polyamide membrane of 5 kDa has been used in the process. Figure 6.11 shows that as the proportion of BN decreases and Cu^{2+} increases in the feed, the retention of BN increases but Cu^{2+} increases. When the concentration of both the contaminants is comparable (Cu^{2+}:BN is 1:0.5), the rejection efficiencies are 91.4 and 82.6% for Cu^{2+} and BN, respectively.

Witek et al.[51] reported MEUF of phenol, p-cresol, xylenol, and Cr^{3+} using CTAB and SDS (not mixed) in a cross-flow setup having a 20 kDa MWCO membrane. Some interesting results are observed in their analysis. In the case of MEUF of the mixtures of phenols (without Cr^{3+}), CTAB produces slightly more rejection than SDS. The removal efficiencies of xylenol at 10 mM surfactant concentration are 80% and 69% using CTAB and SDS, respectively. The presence of Cr^{3+} does not affect the removal efficiency of phenols using SDS. However, removal of Cr^{3+} increases sharply at a low concentration of SDS. For example, the rejection of Cr^{3+} is 94% at a surfactant concentration of 10 mM and remains invariant at higher SDS concentration. Phenols and other alcohols solubilize in the interior core of the micelle with the hydroxyl group attached to the head region of an anionic micelle. In the cationic surfactant, there exists a partial charge delocalization of the π-electrons of the aromatic solutes due to the ion-induced dipole interaction of the cationic surfactant hydrophilic groups, which is responsible for the enhanced removal by cationic surfactant.[52] In the case of anionic surfactant,

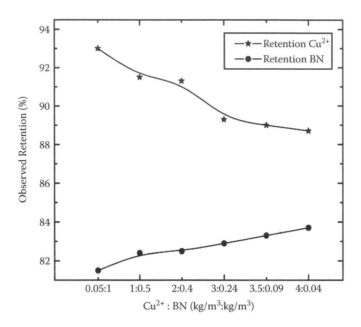

FIGURE 6.11
Retention of Cu^{2+} and BN with varying feed concentrations.

the phenol solubilization is unaffected, whether the countercations neutralize the micellar charge in the Stern layer or the electrical double layer, unless the shape of the micelle is distorted significantly, due to the large amount of electrolytes. Hence the presence of added electrolyte reduces the removal efficiency of contaminants. Therefore, a small concentration of multivalent cations present in the waste stream does not affect the rejection of organic contaminants.[53]

Simultaneous removal of chlorinated hydrocarbons, 1-chlorobenzenes (MCBs) and 1,2-dichlorobenzene (DCBs), and oxy-anionic pollutants, chromate and nitrate, has been investigated using MEUF in the cross-flow mode using a 5 kDa polysulfone (PS) membrane operated at 2 bar TMP.[54] It is evident from Baek and Yang's results that the removal of chromate and nitrate remains unaffected by the presence of the organic contaminants (DCBs and MCBs). In the DCB/CPC system, the addition of nitrate or chromate decreases the removal of DCBs marginally. Addition of MCBs also lowers the removal of DCBs. In a MCB/DCB/CPC system, a decrease in the removal of DCBs is due to the addition of nitrate or chromate. Removal of DCBs was less than 2%. In the case of MCBs, removal is less than 10%. Nitrate and chromate removal decrease slightly, less than 0.2% for chromate and less than 3% for nitrate, in the presence of MCBs or DCBs (refer to Figure 6.12).

A quaternary mixture of Cu^{2+}, PNP, BN, and aniline is selected for pollutant removal using SDS. The compositions are selected such that all the

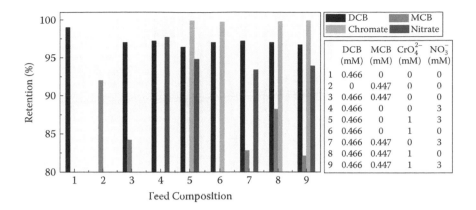

FIGURE 6.12 (See color insert.)
Retention of different solutes using 5 kDa PS membrane in cross-flow mode using 20 mM CPC at 200 kPa.

solutes vary in a lower to higher range. The concentrations (in kg/m³) of solutes are Cu²⁺:BN:PNP:aniline = 0.05:0.04:0.04:0.05, 0.05:0.04:0.04:0.5, etc. Variations of permeate flux for these 11 compositions of mixture are plotted in Figure 6.13. There are three distinct zones in the figure. The first zone is from compositions 1 to 5. The second consists of composition 6, and the third zone is from 7 to 11. In the first zone, Cu²⁺ ion concentration is constant at 0.05 kg/m³ and other organic solute concentrations increase regularly. As a result, flux declines from 18.54 to 14.4 L/m².h. In zone 2, all solute concentrations are between those of zones 1 and 3. Therefore, flux is in between zones

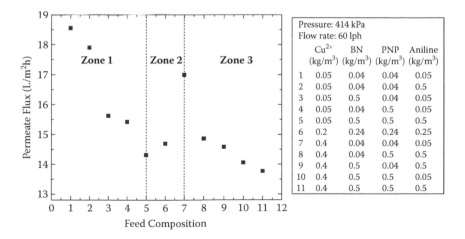

FIGURE 6.13
Effect of feed composition on permeate flux in Cu, BN, PNP, and aniline mixture.

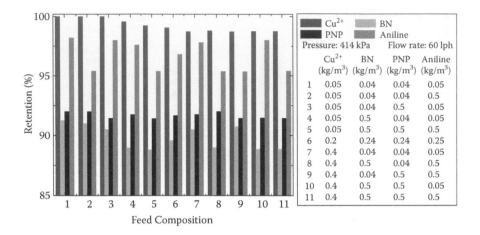

FIGURE 6.14 (See color insert.)
Effect of feed composition on solute retention in Cu, BN, PNP, and aniline mixture (5 kDa PS membrane).

1 and 3, and it is about 15 L/m^2.h. In zone 3, Cu^{2+} concentration is maximum, i.e., 0.4 kg/m^3. All other solute concentrations are increased, and flux therefore decreases from 17 to 14 L/m^2.h (refer to Figure 6.14). About 10% of flux decline is observed in the range of concentrations studied herein. Retention value for copper is the highest, between 97 to 100%. For organic solutes PNP and aniline, retention values are in between 92 and 97%. For BN, it is from 87 to 92%. The retentions of every solute are slightly less than those of pure component systems, due to competitive solubilization, as discussed earlier.

References

1. Arvanitoyannis, I., Eleftheriadis, I., and Tsatsaroni, E. 1989. Influence of pH on adsorption of dye containing effluents with different bentonites. *Chemosphere* 18: 1707–1711.
2. Shu, H.Y., Huang, C.R., and Chang, M.C. 1994. Decolourization of mono-azo dyes in wastewater by advanced oxidation process: A case study of acid red and acid yellow 23. *Chemosphere* 29: 2597–2607.
3. Glaze, W.H., Kang, W.J., and Chapin, D.H. 1987. The chemistry of water treatment process involving ozone, hydrogen peroxide and ultraviolet radiation. *Ozone Sci. Eng.* 9: 335–352.
4. Tang, W.Z., and Huren, A. 1995. UV/TiO$_2$ photocatalytic oxidation of commercial dyes in aqueous solutions. *Chemosphere* 31: 4157–4170.
5. Al-Degs, Y., Khraisheh, M.A.M., Allen, S.J., and Ahmad, M.N.A. 2001. Sorption behavior of cationic and anionic dyes from aqueous solution on different types of activated carbons. *Sep. Sci. Technol.* 36: 91–102.

6. Pollock, M. 1973. Neutralizing dye-house wastes with flue gases and decolorizing with fly ash. *Am. Dyestuff Rep.* 62: 21–23.

7. Chakraborty, S., Purkait, M.K., DasGupta, S., De, S., and Basu, J.K. 2003. Nanofiltration of textile plant effluent for color removal and reduction in COD. *Sep. Purif. Technol.* 31: 141–151.

8. Purkait, M.K., DasGupta, S., and De, S. 2004. Removal of dye from wastewater using micellar-enhanced ultrafiltration and recovery of surfactant. *Sep. Purif. Technol.* 37: 81–92.

9. Purkait, M.K., DasGupta, S., and De, S. 2006. Micellar enhanced ultrafiltration of eosin dye using hexadecyl pyridinium chloride. *J. Hazard. Mater.* B136: 972–977.

10. Purkait, M.K., DasGupta, S., and De, S. 2004. Resistance in series model for micellar enhanced ultrafiltration of eosin dye. *J. Colloid Interf. Sci.* 270: 496–506.

11. Ahmad, A.L., Puasa, S.W., and Zulkali, M.M.D. 2006. Micellar-enhanced ultrafiltration for removal of reactive dyes from an aqueous solution. *Desalination* 191: 153–161.

12. Koyuncu, I. 2002. Reactive dye removal in dye/salt mixtures by nanofiltration membranes containing vinylsulphone dyes: Effect of feed concentration and cross flow velocity. *Desalination* 143: 243–253.

13. Zaghbani, N., Hafiane, A., and Dhahbi, M. 2007. Separation of methylene blue from aqueous solution by micellar enhanced ultrafiltration. *Sep. Purif. Technol.* 55: 117–124.

14. Bielska, M., and Szymanowski, J. 2006. Removal of methylene blue from waste water using micellar enhanced ultrafiltration. *Water Res.* 40: 1027–1033.

15. Poiniak, G., Poiniak, R., and Wilk, K.A. 2009. Removal of dyes by micellar enhanced ultrafiltration. *Chem. Eng. Trans.* 17: 1693–1697.

16. Göktürk, S. 2005. Effect of hydrophobicity on micellar binding of carminic acid. *J. Photochem. Photobiol. A Chem.* 169: 115–121.

17. Zaghbani, N., Hafiane, A., and Dhahbi, M. 2008. Removal of Safranin T from wastewater using micellar enhanced ultrafiltration. *Desalination* 222: 348–356.

18. Kumar, S., Mohanty, K., and Meikap, B.C. 2010. Removal of phenol from dilute aqueous solutions in a multistage bubble column adsorber using activated carbon prepared from *Tamarindus indica* wood. *J. Environ. Prot. Sci.* 4: 1–7.

19. Meikap, B.C., and Rot, G.K. 1997. Removal of phenolic compounds from industrial waste water by semifluidized bed bio-reactor. *J. IPHE India* 3: 54–61.

20. Purkait, M.K., DasGupta, S., and De, S. 2005. Separation of aromatic alcohols using micellar-enhanced ultrafiltration and recovery of surfactant. *J. Membr. Sci.* 250: 47–59.

21. Syamal, M., De, S., and Bhattacharya, P.K. 1997. Phenol solubilization by cetyl pyridinium chloride micelles in micellar-enhanced ultrafiltration. *J. Membr. Sci.* 137: 99–107.

22. Zeng, G.M., Xu, K., Huang, J.-H., Li, X., Fang, Y.-Y., and Qu, Y.-H. 2008. Micellar enhanced ultrafiltration of phenol in synthetic wastewater using polysulfone spiral membrane. *J. Membr. Sci.* 310: 149–160.

23. Sabate, J., Pujola, M., and Llorens, J. 2002. Comparison of polysulfone and ceramic membranes for the separation of phenol in micellar-enhanced ultrafiltration. *J. Colloid Interf. Sci.* 246: 157–163.

24. Talens-Alesson, F.I., Adamczak, H., and Szymanowski, J. 2001. Micellar-enhanced ultrafiltration of phenol by means of oxyethylated fatty acid methyl esters. *J. Membr. Sci.* 192: 155–163.

25. Adamczak, H., Materna, K., Urbanski, R., and Szymanowski, J. 1999. Ultrafiltration of micellar solutions containing phenols. *J. Colloid Interf. Sci.* 218: 359–368.

26. Bielska, M., and Szymanowski, J. 2004. Micellar enhanced ultrafiltration of nitrobenzene and 4-nitrophenol. *J. Membr. Sci.* 243: 273–281.

27. Abraham, M.H., Chadha, H.S., Dixon, J.P., Rafols, C., and Treiner, C. 1995. Hydrogen-bonding. 40. Factors that influence the distribution of solutes between water and sodium dodecyl-sulfate micelles. *J. Chem. Soc. Perkin Trans.* 2: 887–894.

28. Abraham, M.H., Chadha, H.S., Dixon, J.P., Rafols, C., and Treiner, C. 1997. Hydrogen-bonding. 41. Factors that influence the distribution of solutes between water and hexadecylpyridinium chloride micelles. *J. Chem. Soc. Perkin Trans.* 2: 19–24.

29. Gibbs, L.L., Scamehorn, J.F., and Christian, S.D. 1987. Removal of n-alcohols from aqueous streams using micellar-enhanced ultrafiltration. *J. Membr. Sci.* 30: 67–74.

30. Hayase, K., and Hayano, S. 1977. The distribution of higher alcohols in aqueous micellar solutions. *Bull. Chem. Soc. Jpn.* 50: 83–85.

31. Scamehorn, J.F., and Harwell, J.H., eds. 1989. *Surfactant based separation processes.* Surfactant Science Series, vol. 33. New York: Marcel Dekker.

32. Miltner, R.J., Fronk, C.A., and Speth, T.F. 1987. *Removal of alachlor from drinking water.* Springfield, VA: NTIS.

33. Hai-Yan, L.I., Jiu-Hui, Q.U., Zhao, X., and Juan, L.H. 2004. Removal of alachlor from water by catalyzed ozonation in the presence of Fe^{2+}, Mn^{2+}, and humic substances. *J. Environ. Sci. Health B Pest. Food. Contam. Agric. Waste* 39: 791–803.

34. Kim, H.Y., Kim, I.K., Shim, et al. 2006. Removal of alachlor and pretilachlor by laboratory-synthesized zerovalent iron in pesticide formulation solution. *Bull. Environ. Contam. Toxicol.* 77: 826–833.

35. Xiarchos, I., and Doulia, D. 2006. Effect of nonionic surfactants on the solubilization of alachlor. *J. Hazard. Mater.* B136: 882–888.

36. Krogh, K.A., Hallig-Sørensen, B., Mogensen, B.B., and Vejrup, K.V. 2003. Environmental properties and effects of nonionic surfactant adjuvants in pesticides: A review. *Chemosphere* 50: 871–901.

37. Xiarchos, I., and Doulia, D. 2006. Interaction behaviour in ultrafiltration of nonionic surfactant micelles by adsorption. *J. Colloid Interf. Sci.* 299: 102–111.

38. Doulia, D., and Xiarchos, I. 2007. Ultrafiltration of micellar solution of nonionic surfactant with or without alachlor pesticide. *J. Membr. Sci.* 296: 58–64

39. Rogers, V.V., Wickstrom, M., Liber, K., and Mackinnon, M.D. 2002. Acute and subchronic mammalian toxicity of naphthenic acids from oil sands tailings. *Toxicol. Sci.* 66: 347–355.

40. Leung, S.S., MacKinnon, M.D., and Smith, R.E.H. 2003. The ecological effects of naphthenic acids and salts on phytoplankton from the Athabasca oil sands region. *Aqua. Toxicol.* 62: 11–26.

41. Headley, J.V., and McMartin, D.W. 2004. A review of the occurrence and fate of naphthenic acids in aquatic environments. *J. Environ. Sci. Health* 39: 1989–2010.

42. Clement, J.S., and Fedorak, P.M. 2005. A review of the occurrence, analyses, toxicity and biodegradation of naphthenic acids. *Chemosphere* 60: 585–600.

43. Angela, C.S., Mackinnon, M.D., and Fedorak, P.M. 2005. Naphthenic acids in Athabasca oil sands tailings waters are less biodegradable than commercial naphthenic acids. *Environ. Sci. Technol.* 39: 8388–8394.

44. Venkataganesh, B. 2011. Electric field assisted micellar enhanced ultrafiltration for removal of naphthenic acid. M.Tech. thesis, Department of Chemical Engineering, IIT Kharagpur, India.

45. Dunn Jr., R.O., Scamehorn, J.F., and Christian, S.D. 1985. Use of micellar-enhanced ultrafiltration to remove dissolved organics from aqueous stream. *Sep. Sci. Technol.* 20: 257–284.

46. Dunn Jr., R.O., Scamehorn, J.F., and Christian, S.D. 1987. Concentration polarization effects in the use of micellar-enhanced ultrafiltration to remove dissolved organic pollutants. *Sep. Sci. Technol.* 22: 763–789.

47. Pramauro, E., Prevot, A.B., Savarino, P., Viscardi, D., de la Guardia, M., and Cardells, E.P. 1993. Preconcentration of aniline derivatives from aqueous solutions using micellar-enhanced ultrafiltration. *Analyst* 118: 23–27.

48. Materna, K., Goralska, E., Sobczynska, A., and Szymanowski, J. 2004. Recovery of various phenols and phenylamines by micellar enhanced ultrafiltration and cloud point separation. *Green Chem.* 6: 176–182.

49. Jadhav, S.R., Verma, N., Sharma, A., and Bhattacharya, P.K. 2001. Flux and retention analysis during micellar enhanced ultrafiltration for the removal of phenol and aniline. *Sep. Purif. Technol.* 24: 541–557.

50. Das, C., Dasgupta, S., and De, S. 2008. Simultaneous separation of mixture of metal ions and aromatic alcohol using cross flow micellar-enhanced ultrafiltration and recovery of surfactant. *Sep. Sci. Technol.* 43: 71–92.

51. Witek, A., Koltuniewicz, A., Kurczewski, B., Radziejowska, M., and Hatalski, M. 2006. Simultaneous removal of phenols and Cr^{3+} using micellar-enhanced ultrafiltration process. *Desalination* 191: 111–116.

52. Lianos, P., Viriot, M.L., and Zana, R. 1984. Study of the solubilization of aromatic hydrocarbons by aqueous micellar solutions. *J. Phys. Chem.* 88: 1098–1101.

53. Dunn Jr., R.O., Scamehorn, J.F., and Christian, S.D. 1989. Simultaneous removal of dissolved organics and divalent metal cations from water using micellar-enhanced ultrafiltration. *Colloids Surfaces* 35: 49–56.

54. Baek, K., and Yang, J.-W. 2004. Simultaneous removal of chlorinated aromatic hydrocarbons, nitrate, and chromate using micellar-enhanced ultrafiltration. *Chemosphere* 57: 1091–1097.

7

Permeate Flux: Influencing Factors

In any membrane-based separation process, permeate flux is extremely important. It represents the productivity or throughput of the filtration process. Moreover, an accurate estimation of permeate flux helps in appropriate scaling up of the process. Therefore, influence of operating conditions on the permeate flux is of utmost importance, and their effects would be understood properly. During membrane filtration, solutes accumulate on the membrane surface, providing an extra resistance against the solvent flux through it. This is known as concentration polarization.[1] This can be manifested in (1) an increase in osmotic pressure at the membrane-solution interface, and (2) the formation of a gel type, highly viscous layer over the membrane surface. These two effects lead to membrane fouling. The first effect reduces the driving force, i.e., pressure gradient across the membrane, and the second offers an extra resistance in series with hydraulic resistance of the membrane. During micellar-enhanced ultrafiltration (MEUF), the concentration polarization is in the form of a highly viscous but porous gel layer, containing the micelles. This type of fouling can be removed by proper washing of the membrane after the filtration. However, during the filtration, this effect is unavoidable, leading to a decline in the permeate flux or throughput of the system. Thus, one has to investigate the methods to minimize the concentration polarization during MEUF. In this chapter, three methods of fouling reduction have been discussed: application of an external electric field, modification of a membrane surface, and hydrodynamic methods.

7.1 Application of External Electric Field

Use of an electric field becomes extremely important in the case of the filtration of charged solutes. During filtration, the solutes get deposited on the membrane surface. On application of an electric field of opposite polarity (opposite to the particle charge) on the top surface of the cell, the particles are lifted up by electrostatic attraction and move through the electrolytic solution. This reduces the thickness of the gel type layer (composed of charged particles) and increases the permeate flux significantly. An external electric field is used to increase the productivity during ultrafiltration/microfiltration of the protein solution,[2-5] and fractionation of the protein mixture,[6]

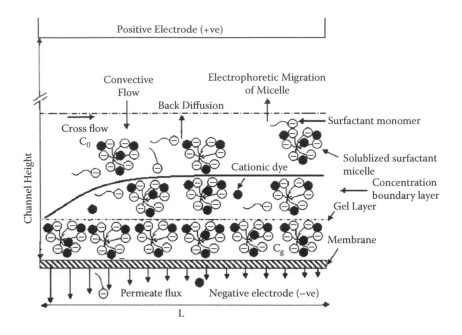

FIGURE 7.1
Schematic of MEUF in presence of external electric field.

clay,[7] pectin solution,[8] fruit juice,[9] etc. Exploiting the charged nature of ionic micelles, an external electric field can be utilized effectively to reduce concentration polarization and increase the permeate flux. Use of an electric field has been successfully investigated during cross-flow microfiltration of a cationic surfactant.[10]

Sarkar et al.[11] have investigated the effects of an external electric field during MEUF of a dye, methylene blue (MB), using sodium dodecyl sulfate (SDS) micelles. Figure 7.1 shows the schematic representation of removal of solute containing micelles from the membrane surface in the presence of an external direct current electric field.

They used 10,000 molecular weight cutoff (MWCO) polyether sulfone membranes in a cross-flow cell under total recycle mode. The concentration range of SDS is 5 to 30 kg/m³, the electric field is in between 0 and 1,000 V/m, transmembrane pressure is varied from 220 to 635 kPa, and cross-flow velocities are in the range of 0.09 to 0.18 m/s. The concentration of MB dye is selected as 0.01, 0.02, 0.03, 0.04, or 0.05 kg/m³. It is observed that the optimum concentration ratio of SDS to MB (concentration expressed in kg/m³) is 400, where the retention of MB is maximum (99%). Table 7.1 presents the variation of permeate flux and observed retention of MB with an electric field strength at an SDS-to-MB ratio of 400, transmembrane pressure drop of 360 kPa, and cross-flow velocity of 0.12 m/s.

TABLE 7.1

Variation of Permeate Flux and Observed Retention of MB with Electric Field Strength

Electric Field (V/m)	Permeate Flux (L/m².h)	Observed Retention of MB (%)
0	44	98
400	47	99
600	51	99
800	57	99
1,000	59	99

It is observed from this table that with an increase in electric field, the permeate flux increases. With an increase in electric field, the gel layer thickness of micelle aggregates on the membrane surface decreases due to electrophoretic movement of the SDS micelle away from the membrane surface. For example, the permeate flux increases 32.5%, with an increase in electric field from 0 to 1,000 V/m. It is also observed from the table that the effect of electric field on the retention of MB is not significant. For example, with the same increment of electric field, the retention of MB changes from 98% to 99%, keeping other operating conditions unchanged. This observation indicates that the retention of dye is entirely dictated by its solubilization only within the micelles. Effects of cross-flow velocity on the permeate flux and retention of MB at the electric field strength E = 800 V/m are also investigated. At this field strength and an SDS-to-MB concentration ratio of 400, with an increase in cross-flow velocity from 0.09 m/s to 0.18 m/s, the permeate flux increases from 52 L/m².h to 58 L/m².h, i.e., by about 11.5%. This is due to enhanced cross-flow velocity; the shear on the gel layer is more, leading to reduction in gel layer thickness. This leads to an increase in permeate flux. On the other hand, the retention of MB does not change significantly. This shows that the solute solubilization in the micelle is independent of the cross-flow velocity in the flow channel. Effects of transmembrane pressure drop on the permeate flux and retention of MB are also investigated. At 800 V/m electric field strength and cross-flow velocity 0.12 m/s, and at optimum concentration of SDS and MB, with an increase in operating pressure from 220 kPa to 360 kPa, the permeate flux increases from 41.5 L/m².h to 61.2 L/m².h, leading to about 48% enhancement of flux. The change on observed retention of MB is marginal.

Venkataganesh[12] has reported the effects of a d.c. electric field on the permeate flux and retention of naphthenic acid (NA) using SDS micelles. A 10,000 MWCO polyether sulfone membrane is used in a cross-flow cell. Two modes of application of electric field, namely, stepwise increase in electric field and continuous electric field at uniform value, are used. In the first mode, the experiment is started at zero electric field at particular operating conditions. After reaching a steady state, the field strength is set to 200 V/m without disturbing other operating conditions. After reaching steady state,

TABLE 7.2

Variation of Permeate Flux with Stepwise Increase in Electric Field

Electric field strength (V/m)	0	200	400	600	800	1,000
Permeate flux (L/m².h)	4.97	5.08	5.15	5.22	5.29	5.40

the next field strength of 400 V/m is set. This process is repeated in steps of 200 V/m up to 1,000 V/m. In the second mode, a fixed electric field is applied from the beginning of the experiment so that the field strength remains constant throughout the experiment. Once the experiment is over, the membrane is cleaned and a new experiment is started with another value of electric field strength. The transmembrane pressure drop is in the range of 276 to 550 kPa and the cross-flow rate is from 40 to 80 LPH. The electric field strength is varied from 200 to 1,000 V/m. Experiments are conducted at a constant NA concentration of 500 mg/L. It is reported that the optimum operating pH is 3.0 and that the SDS concentration is 8 critical micellar concentration (CMC). Effects of a stepwise increase in electric field on the steady-state permeate flux at each electric field are reported in Table 7.2, at $\Delta P = 276$ kPa, cross-flow rate 80 LPH, and other optimum operating conditions.

Table 7.2 clearly shows an increment of permeate flux by 9% when the electric field is increased stepwise from 0 to 1,000 V/m, keeping other operating conditions fixed. The effects of application of electric field in the second mode on the steady-state permeate flux at each electric field are reported in Table 7.3, at $\Delta P = 276$ kPa, cross-flow rate 80 LPH, and other optimum operating conditions.

Table 7.3 clearly shows an increment of permeate flux by 18% when the electric field is increased continuously at a uniform value of 0 to 1,000 V/m, keeping other operating conditions fixed. This is due to the fact that since in mode 2 the electric field is applied from the beginning of the filtration, it will be more effective than a stepwise increase in electric field. On the other hand, the values of observed retention are independent of electric field strength and remain in the range of 94 to 98%, depending on the operating conditions. Thus, application of an external electric field influences the permeate flux significantly by reducing the concentration polarization, but it does not affect the retention of solute solubilization within the micelles.

TABLE 7.3

Variation of Permeate Flux with Mode 2 Increase in Electric Field

Electric field strength (V/m)	0	200	400	600	800	1,000
Permeate flux (L/m².h)	4.97	5.22	5.40	5.58	5.76	5.83

7.2 Surface Modification

Polysulfone (PS) and polyacrylonitrile (PAN) are commonly used polymers for manufacturing ultrafiltration membranes because of their good thermal and chemical stabilities.[13] The polymeric membrane surface is hydrophobic in nature, resulting in severe fouling due to the deposition of the solute particles. A drastic decrease in hydrophobicity and an increase in hydrophilicity can be done by several methods, including chemical treatment, plasma treatment, ion beam irradiation, physical adsorption of modifiers (e.g., surfactants, block copolymers), grafting, etc.

7.2.1 Chemical Treatment and Physical Coating

Membrane performance can be improved by modifying its surface using chemical treatment. Surface modification by chemical treatment using hydrophilicity enhancing agents,[14] proteic acids,[15] is reported in the literature. Dimov and Islam[15] have reported that permeate flux can be enhanced by chemical modification of the microfiltration membrane using a ternary mixture of ethanol-water-inorganic acids. Mukherjee et al.[16] have reported that chemical treatment of polyamide RO membrane with a solution of hydrofluoric acid, fluosilicic acid, and isopropyl alcohol increases both flux and rejection. Membrane performance can also be enhanced by physical coating of the membrane surface. Physical coating of poly(vinylidene fluoride), a polyamide membrane with a specific type of polyether-polyamide block copolymer, results in higher permeate flux than the uncoated membrane.[17]

7.2.2 Plasma Treatment

Among the various surface modification techniques, low-temperature plasma treatment is regarded as the most advantageous one. The PAN surface becomes almost twice as polar after air plasma treatment (increase of polarity from 44 to 85%),[18] and PE even 10 times more polar (increase from 6 to 60%)[19] than unmodified polymeric membranes. For ultrafiltration of an anionic surfactant solution such types of membranes are highly desirable. Techniques for increasing the hydrophobicity or implanting hydrophobic characteristics of the membrane surface are an advanced level of research even today. Various methods like low-temperature plasma treatment,[20] oxygen plasma treatment,[21] carbon dioxide plasma treatment,[22] chemical modifications of the membrane material composition,[23,24] irradiation methods,[25] etc., are used for this purpose. Table 7.4 summarizes the effects of plasma medium on the properties of various membranes. It can be inferred from Table 7.4 that as the polarity of the membrane surface increases, the

TABLE 7.4

Effects of Plasma Treatment on Membrane Properties

Base Material	Plasma Medium	Surface Tension (mN/m)	Polarity (%)	Recovery of Flux after Cleaning	Reduction in Flux	Recovery of Flux after Cleaning	Reduction in Flux
		Surface Properties		Filtration Properties			
				pH = 3		pH = 8	
Polysulfone (PS)	None	45.9	2.0	53.2	82.5	68.4	72.9
	CO_2	61.8	51.8	88.2	67.5	100.0	35.1
	N_2	59.2	49.0	87.0	62.5	89.5	59.8
	NH_3	55.3	47.0	69.8	73.2	72.8	71.1
	N-butyl amine	37.2	18.0	36.21	69.1	48.7	53.3
	N-butyl amine + Argon	44.6	22.9	92.1	54.7	48.9	78.9
	Allyl amine	44.9	24.3	77.6	61.7	86.0	58.3
	Allyl amine + Argon	76.5	52.3	90.3	28.8	100.0	20.0
	Acrylic acid	67.6	63.0	45.1	82.2	76.5	64.3
Poly (phenylene oxide) (PPO)	None	40.6	5.1	113	50	113	42
	Sulfonation with chlorosulfonic acid of PPO membrane modified by amine plasma polymerization (SPPO/ppall)	32.7	53.1				
	Grafting of sodium styrenesulfonate (NaSS) on PPO membrane modified by amine plasma polymerization (GPPO/ppall)	30	36.2	110	34	98	38

hydrophilicity increases. The sulfonated membranes are more effective in MEUF than the unmodified membranes. However, in the case of polyphenylene oxide (PPO) modified membranes, sulfonation with chlorosulfonic acid of a PPO membrane modified by amine plasma polymerization (SPPO/ppall) is more effective than GPPO/ppall. The rejection in the case of SPPO/ppall is more than 90% using a cationic surfactant, while using GPPO/ppall the rejection is less than 50%.[26]

7.2.3 Ion Beam Irradiation

The effect of ion beam irradiation has a positive impact on the membrane performance.[27,28] Ion beam irradiation is the bombardment of a substance with energetic ions. As the ions penetrate the membrane, they lose energy to the membrane polymer, resulting in bond breaking, cross-linking, and formation of volatile molecules. This alters the microstructure of the polymer and morphology of the surface, resulting in a reduction in surface roughness. This reduction in the membrane roughness significantly enhances its performance because smooth membrane surfaces are less prone to fouling than the rough surfaces.

7.2.4 Grafting Polymers

To reduce the fouling resistance, membrane surfaces are modified by grafting polymers.[29] One of the widely used graft polymerization techniques for membrane surface modification is ultraviolet (UV)-assisted photochemical grafting.[30] Plasma-initiated grafting is another method used for membrane surface modification.[31–33] In this method, the membrane is treated by plasma (e.g., argon plasma), and then, post-plasma treatment, grafting of a hydrophilic polymer, is done from the vapor phase. For example, to prepare hydrophilic nanofiltration membranes, low-temperature argon plasma treatment, and subsequent grafting with acrylic acid, has been studied with poly(acrylonitrile) membranes. Redox initiation grafting also has been used for membrane surface modification.[34] The modification leads to increased hydrophilicity, which decreases fouling tendency, while enhancing the permeability and selectivity properties of the membrane. Thermal grafting[35] is also used for surface modification, in which the membrane surface is pretreated to add surface-bound vinyl groups that serve as the anchoring sites for polymer chains that are grown from the surface by thermally activated, free radical graft polymerization. Surface-initiated atom transfer radical polymerization (ATRP) is also used to modify regenerated cellulose UF membranes. ATRP grafting has been used only recently for the surface modification of gas permeation, microfiltration, and pervaporation membranes.[36]

7.3 Hydrodynamic Modifications

By modification of the membrane surface, interaction between the solute and membrane can be made favorable to minimize fouling. But, once the cake or gel type layer is formed, the only alternative left is to change the system hydrodynamics so that mass transfer can be improved. Hydrodynamics in the flow channel can be altered either by the steady-state technique or by imposing instability to the flow. In the steady-state technique, high cross-flow velocity or stirring can be used.

7.3.1 Turbulent Flow

An increase in cross-flow velocity in the flow channel is the simplest way to reduce the concentration polarization and fouling on the membrane surface. With an increase in cross-flow velocity, the membrane surface concentration decreases by the forced convection due to increased turbulence, and therefore diffusional flux from the membrane increases as the concentration gradient from the surface to the bulk increases, leading to the enhancement of permeate flux.[37] But high turbulence may create an axial pressure drop, which may in turn decrease the transmembrane pressure drop and increase the pumping overhead.[38]

7.3.2 Unsteady Flows and Induction of Instabilities

As first noticed by Thomas in 1973,[39] the introduction of instabilities in the flow channel results in an enhancement of mass transfer. Hydrodynamic instabilities can be caused by (1) turbulent promoter, (2) gas sparging, (3) secondary flow, and (4) pulsatile flow.

7.3.2.1 Turbulence Promoter

A variety of turbulence promoters have been investigated. The insertion of rods, wire rings, glass beads, a kenics mixer, doughnut disk baffles, and moving balls in the flow channel has been identified to minimize fouling.[40–43] The function of the turbulence promoter is to create local turbulence in the flow channel by increasing velocity and wall shear rate. Using this technique, permeate flux is found to be augmented several times.

7.3.2.2 Gas Sparging

Gas sparging, injecting gas bubbles into the feed, has recently been identified as an effective technique to enhance the performance of ultrafiltration and the microfiltration membrane.[44–49] The addition of air to the liquid stream increases both the turbulence at the surface of the membrane and the superficial cross-flow velocity within the system, suppressing boundary layer formation, leading to an enhancement of the filtration process. Using this technique, Cui et al.[44] have reported a 250% improvement in flux compared to conventional cross-flow operation for the ultrafiltration of dyed dextran solution. Lee et al.[49] have also reported the use of air slugs to improve the cross-flow filtration of bacterial suspensions.

7.3.2.3 Secondary Flow

Application of dynamic filtration using the rotating cylindrical membrane is found to be useful due to excellent fluid mixing and high shear rate. In a rotating membrane, Taylor vortices are formed in the annular portion due to centrifugal flow instability, which is useful to control the fouling. Using this technique, significant improvement in permeate flux and solute retention has been observed by several investigators.[50-52] High energy consumption for rotating the device, problems in maintenance and replacement of the membrane, and scale formation are the major drawbacks of such a rotating device. To overcome these problems, Dean vortices have been suggested by several researchers.[53,54] Dean vortices are formed by forcing the fluids in a curved channel at a modified Reynolds number (Dean number). These Dean vortices improve the mass transfer coefficient by reducing concentration polarization, leading to an enhancement of permeate flux. Several reports on the use of Dean vortices are available in the literature.[55-59]

7.3.2.4 Pulsatile Flow

Use of the pulsation can be applied to the membrane systems not only in the flow pattern[60] but also in the TMP pulsation.[61] The effect of pulsation leads to disturbances in the medium resulting in dispersion of the deposition of the solute particles, which ultimately increases the permeate flux. Use of pulsatile flow in RO has been investigated by Kennedy et al.[62] The effects of pulsation in the UF and MF processes were studied by Najarian et al.[63] and Finnigan et al.[64] The entire mass transfer boundary layer problem for oscillatory Newtonian laminar flow was solved by Illias and Govind,[65] and the results were compared with the experimental data of Kennedy et al.[62] A combination of pulsations along with baffles has been analyzed, and the results show a significant increase in flux.[66]

References

1. Porter, M.C. 2005. *Handbook of industrial membrane technology*. New Delhi: Crest Publishing House.
2. Sarkar, B., DasGupta, S., and De, S. 2008. Prediction of permeate flux during osmotic pressure controlled electric field enhanced cross flow ultrafiltration. *J. Colloid Interf. Sci.* 319: 236–246.

3. Yukawa, H., Shimura, K., and Maniwa, A. 1983. Characteristics of cross flow electroultrafiltration for colloidal solution of protein. *J. Chem. Eng. Jpn.* 16(3): 246–248.

4. Yukawa, H., Shimura, K., and Maniwa, A. 1983. Cross flow electro-ultrafiltration for colloidal solution of protein. *J. Chem. Eng. Jpn.* 16(4): 305–311.

5. Enevoldsen, A.D., Hansen, E.B., and Jonsson, G. 2007. Electro-ultrafiltration of amylase enzyme: Process design and economy. *Chem. Eng. Sci.* 62: 6716–6725.

6. Sarkar, B., DasGupta, S., and De, S. 2009. Electric field enhanced fractionation of protein mixture using ultrafiltration. *J. Membr. Sci.* 341: 11–20.

7. Moulik, S.P., Cooper, F.C., and Bier, M. 1967. Forced-flow electrophoretic filtration of clay suspensions. *J. Colloid Interf. Sci.* 24: 427–432.

8. Sarkar, B., DasGupta, S., and De, S. 2008. Effect of electric field during gel-layer controlled ultrafiltration of synthetic and fruit juice. *J. Membr. Sci.* 307:268–276.

9. Sarkar, B., DasGupta, S., and De, S. 2009. Flux decline during electric field assisted cross flow ultrafiltration of mosambi (*Citrus sinensis* (L.) Osbeck) juice. *J. Membr. Sci.* 331.75–83.

10. Wakeman, R.J., and Tarleton, E.S. 1987. Membrane fouling prevention in cross-flow microfiltration by the use of electric fields. *Chem. Eng. Sci.* 42: 829–842.

11. Sarkar, B., DasGupta, S., and De, S. 2009. Application of external electric field to enhance the permeate flux during micellar enhanced ultrafiltration. *Sep. Purif. Technol.* 66: 263–272.

12. Venkataganesh, B. 2011. Electric field assisted micellar enhanced ultrafiltration for removal of naphthenic acid. M.Tech. thesis, Department of Chemical Engineering, IIT Kharagpur, India.

13. Wu, S. 1987. *Polymer interface and adhesion.* London: Chapman & Hall.

14. Nystrom, M., and Jarvinen, P. 1991. Modification of ultrafiltration membrane with UV irradiation and hydrophilicity agent. *J. Membr. Sci.* 60: 275–296.

15. Dimov, A., and Islam, M.A. 1990. Hydrophilization of polyethylene membrane. *J. Membr. Sci.* 50:97–100.

16. Mukherjee, D., Kulkarni, A., and Gill, W.N. 1996. Chemical treatment for improved performance of reverse osmosis membranes. *Desalination* 104: 239–249.

17. Louie, J.S., Pinnau, I., Ciobanu, I., Ishida, K.P., Ng, A., and Reinhard, M. 2006. Effects of polyether–polyamide block copolymer coating on performance and fouling of reverse osmosis membranes. *J. Membr. Sci.* 280: 762–770.

18. Bryjak, M., Gancarz, I., Krajciewicz, A., and Pigłowski, J. 1996. Air plasma treatment of polyacrylonitrile porous membrane. *Angew. Makromol. Chem.* 234: 21–29.

19. Bryjak, M., and Garncarz, I. 1994. Plasma treatment of polyethylene ultrafiltration membranes. *Angew. Makromol. Chem.* 219: 117–124.

20. Ulbricht, M., and Belfort, G. 1996. Surface modification of ultrafiltration membranes by low temperature plasma II. Graft polymerization onto polyacrylonitrile and polysulfone. *J. Membr. Sci.* 111: 193–215.

21. Kim, K.S., Lee, K.H., Cho, K., and Park, C.E. 2002. Surface modification of polysulfone ultrafiltration membrane by oxygen plasma treatment. *J. Membr. Sci.* 199:135–145.

22. Wavhal, D.S., and Fisher, E.R. 2005. Modification of poysulphone UF membranes by CO_2 plasma treatment. *Desalination* 172:189–205.

23. Nabe, A., Staude, E., and Belfort, G. 1997. Surface modification of polysulfone ultrafiltration membranes and fouling by BSA solutions. *J. Membr. Sci.* 133:57–72.

24. Steuck, M.J., and Reading, N. 1986. Porous membranes having hydrophilic surface and processes. U.S. Patent 4,618,533.
25. Nystrom, M., and Jarvinen, P. 1991. Modification of polysulfone ultrafiltration membranes with UV irradiation and hydrophilicity increasing agents. *J. Membr. Sci.* 60: 275–296.
26. Poźniak, G., Gancarz, I., and Tylus, W. 2006. Modified poly(phenylene oxide) membranes in ultrafiltration and MEUF of organic compounds. *Desalination* 198: 215–224.
27. Good, K., Escobar, I., Xu, X., Coleman, M., and Ponting, M. 2002. Modification of commercial water treatment membranes by ion beam irradiation. *Desalination* 146:259–264.
28. Chennamsetty, R., Escobar, I., and Xu, X. 2006. Characterization of commercial water treatment membranes modified via ion beam irradiation. *Desalination* 188: 203–212.
29. Tang, C., and Chen, V. 2002. Nanofiltration of textile wastewater for water reuse. *Desalination* 143:11–20.
30. Taniguchi, M., Kilduff, J.E., and Belfort, G. 2003. Low fouling synthetic membranes by UV-assisted graft polymerization: Monomer selection to mitigate fouling by natural organic matter. *J. Membr. Sci.* 222:59–70.
31. Ulbricht, M., and Belfort, G. 1995. Surface modification of ultrafiltration membranes by low-temperature plasma. 1. Treatment of polyacrylonitrile. *J. Appl. Polym. Sci.* 56:325–343.
32. Chen, H., and Belfort, G. 1999. Surface modification of poly (ether sulfone) ultrafiltration membranes by low-temperature plasma-induced graft polymerization. *J. Appl. Polym. Sci.* 72:1699–1711.
33. Wavhal, D.S., and Fisher, E.R. 2002. Hydrophilic modification of polyethersulfone membranes by low temperature plasma-induced graft polymerization. *J. Membr. Sci.* 209: 255–269.
34. Belfer, S., Fainshtain, R., Purinson, Y., Gilron, J., Nyström, M., and Mänttäri, M. Modification of NF membrane properties by in situ redox initiated graft polymerization with hydrophilic monomers. *J. Membr. Sci.* 239: 55–64.
35. Faibish, R.S., and Cohen, Y. 2001. Fouling and rejection behavior of ceramic and polymer-modified ceramic membranes for ultrafiltration of oil-in-water emulsions and microemulsions. *Colloids Surf. A Phys. Eng. Aspects* 191: 27–40.
36. Singh, N., Chen, Z., Tomera, N., Wickramasinghe, S.R., Soice, N., and Husson, S.M. 2008. Modification of regenerated cellulose ultrafiltration membranes by surface-initiated atom transfer radical polymerization. *J. Membr. Sci.* 311:225–234.
37. Blatt, W.F., Dravid, A., Michaels, A.S., and Nelson, L. 1970. Solute polarization and cake formation in membrane ultrafiltration: Causes, consequences and control techniques. In *Membrane science and technology*, ed. J.E. Flinn. New York: Plenum Press 47–97.
38. Belfort, G., Davis, R.H., and Zydney, A.L. 1994. The behavior of suspensions and macromolecular solutions in cross microfiltration. *J. Membr. Sci.* 96:1–58.
39. Thomas, D.G. 1973. Forced convection mass transfer in hyperfiltration at high fluxes. *Ind. Eng. Chem. Fundam.* 12:396–405.
40. Nijarian, S., and Bellhouse, B.J. 1996. Enhanced microfiltration of bovine blood using a tubular membrane with screw-threaded insert and oscillatory flow. *J. Membr. Sci.* 112: 249–261.

41. DaCosta, A.R., Fane, A.G., and Wiley, D.E. 1993. Ultrafiltration of whey protein solution in spacer filled flat channel. *J. Membr. Sci.* 76: 245–254.

42. Mackley, M.R., and Sherman, N.E. 1993. Cake filtration mechanisms in steady and unsteady flows. *J. Membr. Sci.* 77: 113–121.

43. Krstic, D.M., Hoflinger, W., Koris, A.K., and Vatai, G.N. 2007. Energy-saving potential of cross-flow ultrafiltration with inserted static mixer: Application to an oil-in-water emulsion. *Sep. Purif. Technol.* 57: 134–139.

44. Cui, Z.F., and Wright, K.I.T. 1994. Gas-liquid two phase flow ultrafiltration of BSA and dextran solution. *J. Membr. Sci.* 90: 183–189.

45. Cui, Z.F., and Wright, K.I.T. 1996. Flux enhancement with gas sparging in downwards crossflow ultrafiltration: Performance and mechanism. *J. Membr. Sci.* 117: 109–116.

46. Mercier, M., Fonade, C., and Delorme, C.L. 1997. How slug flow can enhance the ultrafiltration flux in mineral tubular membrane. *J. Membr. Sci.* 128: 103–113.

47. Cabassud, C., Laborie, S., and Laine, J.M. 1997. How slug flow can improve ultrafiltration flux in organic hollow fibres. *J. Membr. Sci.* 128: 93–101.

48. Ghosh, R., Li, Q., and Cui, Z.F. 1998. Fractionation of BSA and lysozyme using ultrafiltration: Effect of gas sparging. *AIChE J.* 44: 61–67.

49. Lee, C.K., Chang, W.G., and Ju, Y.H. 1993. Air slugs entrapped cross-flow filtration of bacterial suspensions. *Biotechnol. Bioeng.* 41: 525–530.

50. Kroner, K.H., and Nissinen, V. 1985. Dynamic filtration of microbial suspensions using an annular rotating filter. *J. Membr. Sci.* 36: 85–100.

51. Park, J.Y., Choi, C.K., and Kimb, J.J. 1994. A study on dynamic separation of silica slurry using a rotating membrane filter: Experiments and filtrate fluxes. *J. Membr. Sci.* 97:263–273.

52. Lee, S., and Lueptow, R.M. 2003. Control of scale formation in reverse osmosis by membrane rotation. *Desalination* 155: 131–139.

53. Chung, K.Y., Bates, R., and Belfort, G. 1993. Dean vortices with wall flux in a curved channel membrane system. Effect of vortices on permeation fluxes of suspensions in microporous membrane. *J. Membr. Sci.* 81: 139–150.

54. Brewster, M.G., Chung, K.Y., and Belfort, G. 1993. Dean vortices with wall flux in a curved channel membrane system. A new approach to membrane module design. *J. Membr. Sci.* 81: 127–137.

55. Ghogomu, J.N., Guigui, C., Rouch, J.C., Clifton, M.J., and Aptel, P. 2001. Hollow-fibre membrane module design: Comparison of different curved geometries with Dean vortices. *J. Membr. Sci.* 181: 71–80.

56. Kaur, J., and Agarwal, G.P. 2002. Studies on protein transmission in thin channel flow module: The role of Dean vortices for improving mass transfer. *J. Membr. Sci.* 196: 1–11.

57. Guigui, C., Manno, P., Moulin, P., et al. 1998. The use of Dean vortices in coiled hollow-fibre ultrafiltration membranes for water and wastewater treatment. *Desalination* 118: 73–77.

58. Srinivasan, S., and Tien, C. 1971. Reverse osmosis in a curved tubular membrane duct. *Desalination* 9: 127–139.

59. Moulin, P., Rouch, J.C., Serra, C., Clifton, M.J., and Aptel, P. 1996. Mass transfer improvement by secondary flows: Dean vortices in coiled tubular membranes. *J. Membr. Sci.* 114: 235–244.

60. Winzeler, H.B., and Belfort, G. 1993. Enhanced performance for pressure driven membrane processes: The argument for fluid instabilities. *J. Membr. Sci.* 80:35–47.

61. Rodgers, V.G.J., and Sparks, R.E. 1991. Reduction of membrane fouling in the ultrafiltration of binary protein mixtures. *AIChE J.* 37: 1517–1528.
62. Kennedy, T.J., Merson, R.L., and McCoy, B.J. 1974. Improving permeation flux by pulsed reverse osmosis. *Chem. Eng. Sci.* 29: 1927–1931.
63. Najarian, S., and Bellhouse, B.J. 1996. Effect of liquid pulsation on protein fractionation using ultrafiltration processes. *J. Membr. Sci.* 114: 245–253.
64. Finnigan, S.M., and Howell, J.A. 1989. The effect of pulsatile flow on ultrafiltration fluxes in a baffled tubular membrane system. *Chem. Eng. Res. Des.* 67: 278–282.
65. Illias, S., and Govind, R. 1990. Potential applications of pulsed flow for increasing concentration polarization in ultrafiltration. *Sep. Sci. Technol.* 25: 1307–1324.
66. Howell, J.A., and Finnigan, S.M. 1990. The effect of pulsed flow on ultrafiltration fluxes in a baffled tubular membrane system. *Desalination* 79: 181–202.

8

Recovery of Surfactants

Recovery of a surfactant after its potential application in micellar-enhanced ultrafiltration (MEUF) is extremely important for economic viability of the process. Apart from being expensive, the presence of a surfactant in the retentate stream is also another source of pollution. There are two possible methods of treatment that involve precipitation of the surfactant either by chemical treatment[1] or by lowering the temperature below the Krafft point.[2] Also there are conventional methods for recovering the surfactant, like vacuum, air or steam stripping, liquid-liquid extraction, etc.[3,4] The method of precipitating the surfactant leaving the organic/inorganic contaminants suspended/dissolved in water is most effective since the surfactant can easily be separated by gravity, ordinary filtration, or centrifugation. A schematic of the recovery process is outlined in Figure 8.1.

Precipitation occurs either by addition of an ion of opposite charge in the solution or by temperature manipulation of the process. If the pollutants present in the solution in organic phase are soluble in water, they will be dissolved in water during the filtration or recovery of the surfactants. On the contrary, if the organics are sparingly soluble in water, they will tend to float on top of the solution once the surfactant molecules are detached from it. However, if the surfactant is sparingly soluble in water, the organic phase tends to precipitate down at the bottom as a liquid phase, trapped in the solid phase surfactant precipitate.[4] The process of precipitation can be categorized as described below.

8.1 Recovery of Anionic Surfactant

Monovalent and multivalent counterions can be added to precipitate the surfactant present in the solution. The surfactant molecule dissociates into respective ions in water. Thus, the simplest means is to use the associated ion that is present in the surfactant to cause precipitation due to a common ion effect. For example:

$$NaC_{12}SO_4 \text{ (s)} \rightleftharpoons C_{12}SO_4^- \text{ (aq.)} + Na^+ \text{ (aq.)}$$

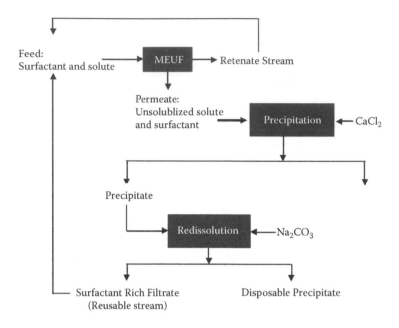

FIGURE 8.1 (See color insert.)
Schematic of a typical anionic surfactant recovery process along with MEUF.

On addition of NaCl, Na^+ ions are in excess, and hence the above reaction shifts more toward the solid SDS, resulting in precipitation to maintain equilibrium. However, using a multivalent ion in slight excess of the stoichiometric proportion is very effective in precipitation. Since the number of surfactant units precipitated per mole of multivalent ion is more than for a monovalent ion, use of a divalent or trivalent salt is more favorable economically. The solubility of ionic surfactant decreases sharply with the magnitude of the charge of the counterion with which it is precipitating.[5,6] But, it has certain disadvantages also. Very fine or colloidal size precipitate is obtained by use of multivalent ions contrary to large size crystals with monovalent ions, which are easy to separate by means of gravity settling or coarse filtration.[4] The effect of a multivalent counterion on the precipitation of SDS is presented in Figure 8.2.

This figure shows that on increasing the ion-to-surfactant ratio beyond the stoichiometric value, SDS precipitates to a large extent. The percentage of SDS precipitation is more for than calcium. However, Al^{3+} being trivalent, it is required in a small quantity and becomes economically viable. One strange feature noticeable from the plot is that the percentage precipitation of SDS passes through a maximum in the case of aluminum, which indicates that the apparent solubility of the SDS is not reduced by adding a trivalent counterion, and the solubility relationship is not simply based on the concentration, as in the case of calcium. Also, aluminum tends to

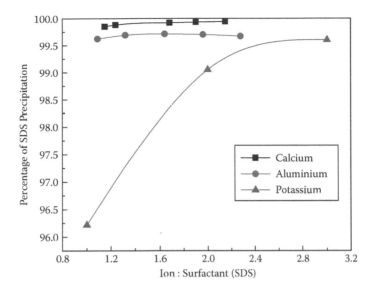

FIGURE 8.2 (See color insert.)
Variation of precipitation of SDS with changing ion/surfactant ratio.

form complexes, thereby requiring a variation of pH to obtain the trivalent aluminum fraction in a maximum amount. For practical purposes, use of calcium is much more feasible than aluminum. To recover the surfactant back into the original form, di- or trivalent complexes have to be treated with their respective similar ion containing carbonate/bicarbonate salt. For example, sodium carbonate is added with the slurry of calcium dodecyl sulfate/aluminum dodecyl sulfate to obtain sodium dodecyl sulfate. Calcium carbonate, thus being insoluble in water, can be filtered out of the SDS solution.

$$Ca(C_{12}SO_4)_2 + Na_2CO_3 \longrightarrow CaCO_3 \downarrow + 2NaC_{12}SO_4$$

$$Al(C_{12}SO_4)_3 + Na_2CO_3 \longrightarrow Al_2(CO_3)_3 \downarrow + 3NaC_{12}SO_4$$

Figure 8.3 shows that when the sodium carbonate is used above the stoichiometric level, SDS dissolution increases considerably.

A maximum of 95% recovery of SDS can be achieved by using the calcium precipitation scheme, while 93% is achieved for aluminum. Since the precipitations using calcium and aluminum are almost quantitatively comparable, the ultimate concept lies in the redissolution step.

Recovery of SDS from retentate solution by adding multivalent counterions such as Ca^{2+} in slight excess of the stoichiometric value can cause precipitation of a high proportion of surfactant. The removal efficiencies of

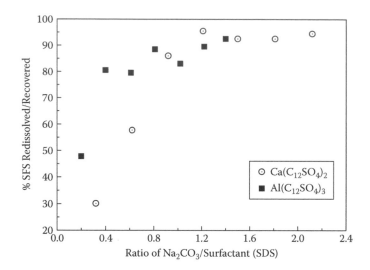

FIGURE 8.3
Percentage of SDS redissolved with changing ratio of Na_2CO_3 to surfactant.

metal ions with reclaimed (recovered) SDS are only 50 to 58%. Precipitating SDS changes the characteristics of micelles so that reuse efficiency is not high enough.

Chelation or acidification followed by ultrafiltration is used for enhanced pollutant removal. It can also be used to recover SDS from a retentate stream via electrostatic interaction. In one such study by Li et al.,[7] for MEUF of Zn^{2+}, the chelating agent EDTA was used for enhanced removal and also for surfactant recovery. It can be observed from Figure 8.4 that the surfactant recovered decreases with pH. This can be explained as, at low pH, a large amount of H^+ ions in solution are close to the surface of SDS micelles with negative charge, and place themselves into the gap between head groups of ions, leading to the charges of micelles shielded. The Stern layer on the micelles' surface is compressed. As a result, an electrostatic repulsive force between head groups of ions is reduced, which accelerates to form more micelles.

In the case of acidulation followed by UF, the recovery of SDS first increases and then decreases with pH. The highest point is 66% with H_2SO_4 at pH 3.0, which is less than the maximum recovery ratio with EDTA. According to the above discussion, at low pH, a large amount of H^+ ions in solution neutralize charges of micelles. As a result, an electrostatic repulsive force is reduced so that the surfactant forms new micelles easily. Simultaneously, in strong acidic condition, hydrolysis of SDS is favored. Another aspect of acidulation is that the surfactant recovery is independent of types of acid used (for, e.g., HCl or HNO_3), at a particular pH.

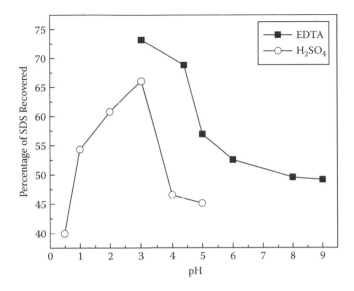

FIGURE 8.4
SDS (feed concentration 3 CMC) recovery using chelating agent EDTA and sulfuric acid (3 M); 6 kDa PES HF membrane is used for removal of Zn^{2+} (feed concentration 100 ppm), TMP 150 kPa.

8.2 Recovery of Cationic Surfactant

Recovery of a cationic surfactant is similar to recovery of an anionic surfactant using a monovalent or multivalent counterion for precipitation. For example, potassium iodide (KI) is used for precipitation of CPC from an effluent stream.[1] The degree of precipitation is highly dependent on the (1) pH of the medium, (2) degree of dissociation of precipitating agent, and (3) temperature. The solubility of CPC in the presence of other anions like SO_4^{2-}, Cl^-, Br^-, and CO_3^- is so high that it does not precipitate even at neutral pH.[8] On the other hand, cetylpyridium iodide (CPI) is highly insoluble at room temperature and neutral pH. Cetylpyridium iodide is obtained by reacting CPC with KI. The reaction scheme is presented in Figure 8.5.

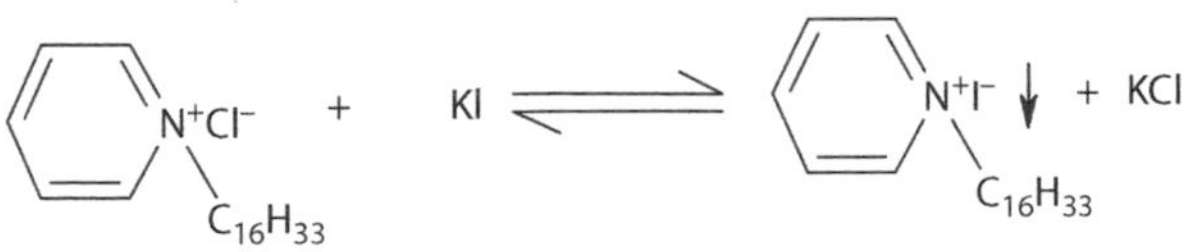

FIGURE 8.5
Precipitation reaction of CPC using KI.

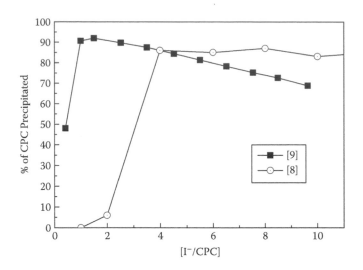

FIGURE 8.6
Variation of CPC precipitated with respect to KI/CPC.

Figure 8.6 shows a comparative study of the two cases by Purkait et al.[1] and Vanjara et al.[8] The fraction of CPC that is precipitated attains a maximum limit of around 90% in the region of the KI/CPC ratio of 2 to 4. The discrepancies in the experimental data may be due to the fact that the data from Purkait et al.[1] are for the permeate of ultrafiltration of the eosin dye and the data of Vanjara et al.[8] are for freshly prepared CPC solution. There may be a presence of an interacting effect due to the presence of other counterions in the solution.

Similar to the recovery of an anionic surfactant, the cationic surfactant needs to be redissolved in the salt solution of a similar ion to recover the original surfactant. In the case of a CPI precipitate, it is redissolved in cupric chloride solution (see Figure 8.7), which enables the formation of CPC that is readily soluble in water, and insoluble cupric iodide is removed by ordinary gravity settling or coarse filtration.

Figure 8.8 shows a comparative analysis of the recovered CPC by Purkait et al.[1] and Vanjara et al.[8] It shows that the percentage recovery can be close to 90% when the ratio of $CuCl_2$ to CPI is between 5 to 7. Further addition of $CuCl_2$ does not improve the percentage recovery of CPC.

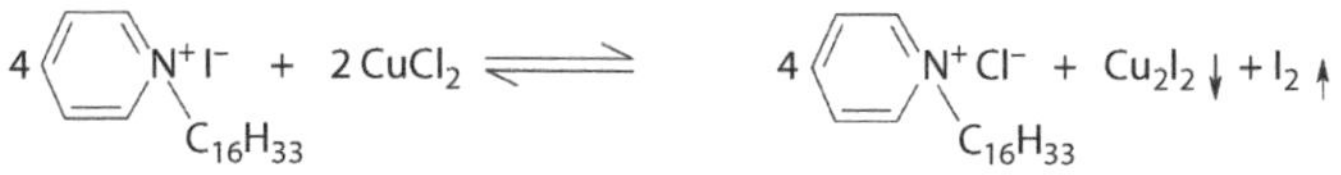

$$4 \left\langle \bigcirc \right\rangle N^+ I^- + 2\,CuCl_2 \;\rightleftharpoons\; 4 \left\langle \bigcirc \right\rangle N^+ Cl^- + Cu_2I_2 \downarrow + I_2 \uparrow$$
$$\quad\quad C_{16}H_{33} \quad\quad\quad\quad\quad\quad\quad\quad\quad\quad C_{16}H_{33}$$

FIGURE 8.7
Redissolution of CPI using $CuCl_2$.

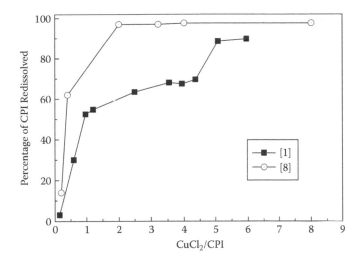

FIGURE 8.8
Percentage of CPI redissolved with changing $CuCl_2$/CPC ratio.

References

1. Purkait, M.K., DasGupta, S., and De, S. 2004. Removal of dye from wastewater using micellar-enhanced ultrafiltration and recovery of surfactant. *Sep. Purf. Technol.* 37. 81–92.
2. Wu, B., Christian, S.D., and Scamehorn, J.F. 1998. Recovery of surfactant from MEUF using a precipitation process. *Progr. Colloid Polym. Sci.*109: 60–73.
3. Roberts, B. 1992. The use of micellar solutions for novel separation techniques. PhD diss., Department of Chemical Engineering, University of Oklahoma.
4. Scamehorn, J.F., and Harwell, J.H., eds. 1989. *Surfactant based separation processes*. Surfactant Science Series, vol. 33. New York: Marcel Dekker.
5. Bozic, J., Krznaric, I., and Kallay, N. 1979. Precipitation and micellization of silver, copper and lanthanum dodecyl sulphates in aqueous media. *Colloid Polym. Sci.* 257: 201–205.
6. Stellner, K.L., and Scamehorn, J.F. 1989. Hardness tolerance of anionic surfactant solutions. 2. Effect of added nonionic surfactant. *Langmuir* 5: 77–84.
7. Li, X., Zenga, G.-M., Huang, J.-H., et al. 2009. Recovery and reuse of surfactant SDS from a MEUF retentate containing Cd^{2+} or Zn^{2+} by ultrafiltration. *J. Membr. Sci.* 337: 92–97.
8. Vanjara, A.K., and Dixit, S.G. 1996. Recovery of cationic surfactant using precipitation method. *Sep. Technol.* 6: 91–93.
9. Purkait, M.K., DasGupta, S., and De, S. 2005. Separation of aromatic alcohols using micellar-enhanced ultrafiltration and recovery of surfactant. *J. Membr. Sci.* 250: 47–59.

9

Other Applications of Micellar-Enhanced Ultrafiltration

Apart from removal of organic and inorganic contaminants from an aqueous stream, there are some other useful applications of micellar-enhanced ultrafiltration (MEUF). The phenomenon of micelle solubilization can be effectively used for separation of bioactive active compounds or separation of the stereo conformers of a particular molecule. Recovery of precious metals and preconcentration of metal ions are also significant comparing the economy of the product obtained and the efficiency of the process. Surfactant-based membrane separation processes are not only cost-effective but also environment friendly.

9.1 Recovery of Precious Metal Ions

Recovery of precious metal ions from the dilute stream is a relevant application of MEUF. Pramauro et al.[1] have reported the recovery of uranium(VI) (U^{VI}) from its dilute acidic mixture using ligand-assisted MEUF. Selective separation of uranium is achieved in the presence of radioactive strontium (Sr^{II}) and other metal ions, like cadmium (Cd^{II}). The ligands used are derivatives of 4-amminosalicylic acid (PAS) and 1-(2-pyridylazo)-2-naphthol (PAN). The principle of separation is that U^{VI} selectively is bound to these ligands, which are solubilized in the nonionic micelles of Triton X-100. Thus, these micelles with bound ligand-containing metal ions are removed by a 10,000 molecular cutoff ultrafiltration (UF) membrane. UF experiments with the surfactant and metal ions only (without ligands) show the retention of these analytes in the order of 3 to 5%. This indicates the positive role of ligands. The ligand-to-metal ratio is kept at 10:1. The ligand and surfactant concentrations are 7×10^{-4} M and 0.02 M, respectively. More than 90% retention of U^{VI} is observed in the pH range of 5 to 6. The corresponding retentions of Sr^{II} and Cd^{II} are below 20%. In a uranyl-containing binary mixture, the experiments show that only 4% of an initial concentration of Sr^{II} is in the retentate after four successive UF steps. In the case of a U^{VI}-Cd^{II} mixture, better results are obtained. Only 1% of initial Cd^{II} is retained. Investigations using a ternary mixture of U^{VI}-Cd^{II}-Sr^{II} indicate 95% retention of U^{VI}, 39% Sr^{II}, and 48% Cd^{II} at pH 3 using PAN-assisted MEUF.

Thus, recovery of uranium in trace amounts can be achieved using MEUF by moderately cheap surfactants and fewer amounts of suitable ligands obtained from common chelating materials in comparison to toxic organic solvents in a liquid-liquid extraction process. This example has possible practical applications in the management of nuclear waste and during decommissioning treatments.

Ghezzi et al.[2] have reported a successful application of MEUF to recover palladium (Pd) from acidic solution. A cationic surfactant dodecyl-trimethylammonium chloride (DTAC) is used for this purpose. The critical micellar concentration (CMC) of DTDC is 0.01 M. A regenerated cellulose membrane of 3,000 molecular weight cutoff (MWCO) is used. Generally, an acidic environment of Pd is maintained to prevent the insoluble metal hydroxides. Only 9% retention of metal is observed at pH 3 without the surfactants. For 0.01 M surfactant, 96% retention of Pd is reported for its initial concentration in the range of 1.3 to 8.5×10^{-5} M. In the range of pH from 2 to 6.5, the retention of Pd increases from 96 to 99%. The Pd bound retentate phase is treated with a suitable volume of stripping solution containing NaCl, Na_2SO_4, and $MgCl_2$. The cations present in the stripping solution displace Pd and a second UF is used to recover Pd in the permeate stream. A maximum 40% recovery of Pd is reported at pH 3.5 with $MgCl_2$, with its concentration 0.8 M. A 66% recovery of Pd is attained by adding various volumes of stripping solution in a multistage manner. It may be mentioned that in the case of recovery of precious metal ions, surfactant recovery is not controlling for economic feasibility.

Gwicana et al.[3] have reported the use of a cationic surfactant for the recovery of platinum group metal ions. Cetylpyridinium chloride (CPC) is used with a polysulfone membrane in a hollow fiber module. More than 90% retention of palladium and platinum is observed. However, detailed experimental conditions and operating conditions are not available in this reference.

Recovery of gold using MEUF is reported by Akita et al.[4] Nonionic surfactant polyoxyethylene nonyl phenyl ethers (PONPEs) with average ethylene oxide numbers 10, 15, and 20 are used. Two cellulose acetate membranes of molecular weight cutoff 3,000 and 10,000 are used. A fixed concentration of gold chloride in the feed is studied (5×10^{-4} M). Almost 100% retention of gold is reported for a PONPE concentration of 0.05 M. The ethylene oxide number has some effects on the gold retention. Retention increases from about 92% to 95% as the ethylene oxide number increases from 10 to 20. Although the effect on retention is marginal in the case of using various membranes, the effects are significant for permeate flux. Permeate flux is increased by almost 50% when a 10,000 cutoff membrane is used compared to a 3,000 cutoff membrane under the same operating conditions. However, the addition of electrolytes like NaCl has an insignificant effect on retention and permeate flux. It is also reported that there is significant selective retention of gold by PONPE compared to other cations, like platinum, palladium, zinc, copper, and iron. This is established by comparing with cationic charged surfactants like CPC. It is shown that the precious metal ions Au^{III}, Pd^{II}, and Pt^{IV} exist in anionic form in dilute acidic solution. In the presence of cationic surfactants,

TABLE 9.1

Salient Points for Recovery of Precious Metals Using MEUF

Precious Metal and Its Concentration	Surfactant and Its Concentration	Membrane System and Operating Conditions	Recovery	Reference
Uranium (VI) (0.025–0.1 mM)	Nonionic surfactant Triton X-100 Ligand: PAS-C_n, where n = 2, 4, 8	10,000 MWCO, cellulose acetate, ΔP = 300 kPa, stirred batch cell Optimum range of operating conditions: pH: 5–6; Surfactant concentration: 10–25 g/L; Ligand concentration: 0.2–0.3 g/L	90%	1
Palladium(II)	Cationic surfactant dodecyltrimethyl-ammonium chloride (DTAC)	3,000 MWCO, regenerated cellulose acetate membrane, ΔP = 300 kPa, stirred batch cell Optimum operating conditions: Metal concentration: 0.013–0.085 mM; Surfactant concentration: 0.03 M; pH: 6.5	99% in micelle phase; metal recovery in stripping phase by 0.8 M $MgCl_2$, pH 3.5: 66%	2
Palladium and platinum	Cetylpyridinium chloride (CPC)	Hollow fiber cross-flow cell using polysulfone membrane	>90%	3
Gold(III)	Nonionic surfactant polyoxyethylene nonyl phenyl ethers (PONPEs) with average ethylene oxide (EO) number 10, 15, and 20	10,000 MWCO cellulose acetate membrane; stirred batch cell; ΔP = 50–300 kPa; stirrer speed: 200 rpm Optimum conditions: Surfactant (EO:20) concentration: 0.05 M; Metal concentration: 0.5 mM	95%	4

retention of these is in the order of 99 to 100% without selectivity. More than 99% surfactant retention is observed, indicating minor leakage of the surfactant to the permeate, indicating insignificant loss of surfactant.

Salient points of various studies related to the recovery of precious metals using MEUF are presented in Table 9.1.

9.2 Recovery of Bioactive Compounds

Amino acids constitute the raw materials in many industries, including pharmaceuticals, chemicals, food, cosmetics, etc.[5,6] These are generally produced by microbial fermentation of molasses or protein hydrolysis.[7,8] In the fermentation broth, the concentration of these amino acids is typically extremely low.[9] They have to be separated from the substrate or by-products. Ion exchange, chromatographic techniques, etc., are used for such purposes.[10] However, these techniques have limitations in terms of production scale, ease, and cost involved. Membrane-based processes can offer an attractive alternative in this regard. They have a distinct advantage of large-scale operation, lower energy requirement, easy to operate and maintain, processing at room temperature, etc.[11] Escudero et al.[9] have investigated the possibility of MEUF for the recovery of α-phenylglycine from its dilute solution. MEUF becomes economically viable when a costly chemical is recovered from its dilute solution. They have undertaken a systematic study by conducting centrifugal ultrafiltration experiments to identify the appropriate molecular cutoff membrane, surfactant-to-amino acid ratio, operating pH, etc. It is observed that up to 85% extraction of α-phenylglycine is obtained for ionic surfactants. For sodium dodecyl benzene sulfonate (SDBS), the pH is 1.5, and at pH 11.2, cetyltrimethylammonium bromide (CTAB) is most effective. Nonionic surfactants, like Tween 80 and amphoteric surfactant SB3-12, are not at all useful. Among 5, 10, and 30 kDa membranes, the 5 kDa membrane is identified to be the best performer. It is observed that extraction of amino acid is marginally better for CTAB, and moreover, the retention of SDBS is less, indicating more SDBS is permeated. Therefore, CTAB is selected as the surfactant and about 50 to 80 mM is the range of optimum surfactant concentration. In the cross-flow experiments, beyond a 350 kPa transmembrane pressure drop, the change in percentage extraction of α-phenylglycine (85%) and retention of the surfactant (about 90%) at 50 mM concentration of surfactant is marginal. For experiments under continuous mode of operation, an almost constant permeate flux of 60 kg/m²h is maintained, even after 28 hours; however, the percentage recovery of surfactant is almost constant at 80%, but the extraction of amino acid drops from 70% to 42% over 28 hours of operation.

Thuringiensin is a low molecular weight (molecular weight 701) insecticidal toxin.[12] Apart from being an insecticide, it is effective in influencing the growth of adults, fertility of flies, maturation of larva, etc.[13] A traditional method to extract thuringiensin involves adsorption of fermentation liquor by ion exchange resin,[14] evaporation at 50°C,[15] salt precipitation,[16] etc. But, these methods lead to a low recovery up to 15%. MEUF has been reported to be effective in this regard.[17] The efficacy of cellulose acetate membrane of various molecular weight cutoffs, namely, 3, 10, and 30 kDa, is investigated in an Amicon stirred batch cell, using cetylpyridinium chloride (CPC) as the surfactant. Cationic CPC is selected because the analyte thuringiensin is anionic in nature. The process involves two steps of MEUF. In the first step, the fermentation broth mixed with CPC micelles is ultrafiltered using 10 and 30 kDa membranes. In the second step, the permeate of the first step is again mixed with CPC solution to recover the permeate thuringiensin using a 3 kDa membrane. It is observed that performance of the 10 kDa membrane is better. In the stirred batch cell, effects of operating conditions, like surfactant concentration, pH, and salt, are investigated and optimized. For scaling up, the two-step MEUF system is carried out at optimized conditions in hollow fiber membrane systems under optimized operating conditions. The ultrafiltration is carried out in both steps up to a retention of 20% feed volume. It is observed from the stirred batch data that the addition of salt is detrimental to the recovery of the analyte, as it neutralizes the ionic character of CPC micelles, lowering the solubilization capacity of the micelles. A maximum 68% recovery is reported without addition of salt, and it is reduced to 15% in the presence of 6% (w/v) salt. The effect of pH in the range of 7 to 9 is not at all significant. The concentration of surfactant is important. It is observed that the recovery of thuringiensin increases with surfactant concentration, and beyond 4% (w/v) the recovery is invariant at about 80%. Between 10 and 30 kDa membrane, the recovery is better in lower cutoff membranes. Temperature in the range of 25 to 45°C has little effect on the performance of MEUF. Experiments in the hollow fiber scaled-up module under optimized conditions indicate that more than 90% recovery of thuringiensin is possible.

Investigations pertaining to recovery of bioactive compounds using MEUF are summarized in Table 9.2.

TABLE 9.2

Summary of Recovery of Bioactive Compounds Using MEUF

Bioactive Compound and Its Concentration	Surfactant and Its Concentration	Membrane System and Operating Conditions	Recovery	Reference
Amino acid, α- phenylglycine (1,000–2,000 mg/L)	Anionic surfactant sodium dodecyl benzene sulfonate (SDBS) and cationic surfactant CTAB; SDBS Best concentration: 80 mM	5,000 MWCO, ceramic membrane, $\Delta P = 400$ kPa; cross-flow cell pH: 11.2 with CTAB and 1.5 for SDBS; 30°C; Cross-flow velocity: 3.5 m/s	85%	9
Thuringiensin	Cationic surfactant CPC Best concentration: 4% (w/v)	Optimum membrane: 10,000 MWCO, cellulose acetate membrane, $\Delta P = 207$ kPa, stirred cell pH: 9.0	88% in first step MEUF and 93% in second step MEUF	17

9.3 Enantioselective Micelles for Separation of Racemic Mixture

Enantiopure compounds have potential applications in pharmaceutical, food, and agro industries. Enantiomers have different biological activities, and hence one enantiomer can have more desired properties than the other. For example, fluazifop butyl has two enantiomers, namely, *R* and *S*. The *R* enantiomer has herbicide properties; the *S* enantiomer is inactive.[18] An *S,S* enantiomer of ethambutol arrests growth of tuberculosis, but an *R,R* enantiomer leads to blindness.[18] These compounds are produced via an asymmetric synthesis route. Thus, it is economically viable to produce these enantiomers in a mixture form of both varieties, known as a racemic mixture, and then, using an appropriate process, the compounds are separated by exploiting their different properties, for example, ligand exchange chromatography for separation of amino acid enantiomers.[18] One of the enantiomers in the racemic mixture has better affinity to a particular ligand, and thus the separation occurs during a chromatography process. Recently, several studies have appeared in the literature based on the concept of ligand-modified micellar-enhanced ultrafiltration for separation of a racemic nixture.[18–21] Separation of phenylalanine (Phe) enantiomers by MEUF is reported by Overdevest et al.[19] In this study, the chiral selectors used are cholesteryl-L-glutamate (CLG) and copper(II). The enantiomers of Phe are D and L. CLG makes a chelating complex with Cu and the D-Phe enantiomer. This complex is preferred over the complex of CLG with Cu and L-Phe. The complex, being hydrophobic in nature, is solubilized within the micelles of nonionic surfactants. These micelles, being larger in size, are separated by the ultrafiltration membrane and thus affect the separation between D and L forms of enantiomers. Overdevest et al.[20] use nonyl phenyl polyoxyethylene ether (NNP10) as the nonionic surfactant. Regenerated cellulose membranes with 3 and 10 kDa molecular cutoffs are used for conducting ultrafiltration experiments. The complexation is carried out in the pH range of 5 to 12. It is observed that pH 9 is an optimum condition where the complexation occurring is maximum. Enantioselectivity decreases with pH. After MEUF, regeneration of enantioselective micelles is carried out by decomplexation at lower pH values. At lower pH (2 to 4), an induced positive charge provides a charge repulsion between CLG, Cu, and the enantiomer. It takes almost 10 hours to regenerate the enantioselective micelles completely at pH 3. A multistage filtration system is finally recommended for 99% separation of enantiomers. Kinetic models for complexation and decomplexation are proposed and examined.[20] A physical model including the complexation kinetics for a cascaded MEUF system for bulk operation to separate the racemic mixture of D and L-Phe is also demonstrated by Overdevest et al.[21] A summary of highlights of studies involving enantioselective micelles is presented in Table 9.3.

TABLE 9.3

Highlights of Enantioselective Micelles during MEUF

Composition of Racemic Mixture	Surfactant with Concentration	Membrane System and Operating Conditions	Selectivity	Reference
Amino acid derivatives D-phenylalanine and L-phenylalanine (0.15 mM each)	Nonionic surfactant nonyl phenyl polyoxyethylene ether (NNP10) Concentration: 7.8 mM Selector compound: Mixture of cholesteryl-L-glutamate (CLG) and Cu^{II} Concentration: 0.3 mM each, 0.1 M KCl	10,000 MWCO, cellulose acetate membrane, $\Delta P =$ 300 kPa, batch cell pH: 9 for complexation of D-phenylanaline with selector compound	Selectivity of D to L form = 4.2	18–20

9.4 Preconcentration Applications

The properties of solubilization of analytes to the micellar aggregates by hydrophobic, electrostatic, and other specific interactions make the analysis of metal ions or toxic organics present in the aqueous stream easier and accurate.[22] The major advantages of this are: (1) the use of a large volume of volatile, toxic, and flammable organic solvents can be avoided; (2) a high preconcentration factor can be achieved, leading to increased sensitivity and a lower detection limit; and (3) variation in the initial concentration of analyte does not alter the extraction efficiency, as surfactant micelles have high preconcentration capacity.[22,23] The principle involved is that the analytes are solubilized in the suitable surfactant micelles (if required, specific ligands are also used) and these aggregates, being larger in size, can easily be removed by an ultrafiltration membrane. The trace amount of analytes and surfactants (at the concentration around CMC) leaks out through the membrane in the permeate stream. If biodegradable surfactants are used, their presence in the permeate does not pose any environmental pollution.

Trace amounts of carcinogenic metal ions, like lead, cadmium, etc., and organics, like amines, benzene, etc., in aqueous streams are toxic. Determination of their concentrations in the samples requires a time-consuming liquid-liquid extraction process for enrichment. An equilibrium-governed extraction process becomes slower at lower concentration of analyte due to a lower driving force. Next, analysis by gas chromatography (GC) or high-pressure liquid chromatography (HPLC) follows. Both methods involve a large number of protocols, and the accuracy of analysis depends on the detection modes. Generally, the accuracy of detection is lower for extremely dilute solution. Thus, an appropriate preconcentration step is required for the analyte for further analysis, and it can be achieved easily by MEUF.[22,23]

Preconcentration of aniline derivatives from an aqueous solution using MEUF is reported by Pramauro et al.[24] Anionic sodium dodecyl sulfate (SDS) and cationic hexadecyltrimethylammonium bromide (HTAB) are used as surfactants. MEUF experiments are conducted in a stirred batch cell using 10 kDa molecular weight cutoff membrane at 300 kPa transmembrane pressure drop. About 70 to 100 ml of solution containing the analytes in ppb level of concentration and the surfactant solution just above its CMC are selected. UF is continued until the retentate volume is reduced to 5% of the original volume. Concentrated retentate (trapped with the analytes) is then analyzed by HPLC using the surfactant solution as the eluting medium to minimize the interference effects. At normal pH of occurrence of aniline derivatives (8 to 8.5), apart from a few out of 20 derivatives studied, most have an extremely high retention of 60 to 100% using both surfactants. In fact, HTAB shows better performance than SDS. Lower pH (2.0 to 3.0) for SDS and higher pH (7.0 to 8.0) for HTAB show extremely good binding of analytes to the micelles. The effect of ionic strength for both surfactants is marginal at higher pH. For an SDS-aniline system, retention is lowered at higher ionic strength for more polar solutes like aniline, 4-methylaniline, and 4-isopropylaniline due to neutralization of the micellar charge. On the other hand, more hydrophobic compounds like alkylanilines have little effect on solution ionic strength because for these compounds, hydrophobic solubilization rather than electrostatic interaction is the main mechanism of solute solubilization. Preconcentration needs to be conducted just above the CMC of the surfactants to avoid high viscosity of retentate solution or the Krafft point in the case of ionic surfactants. For all the compounds, 14 to 20 times concentration is achieved. In the analysis by HPLC, a surfactant-added eluent is used as the medium. This results in excellent reproducibility and stability of the peaks in chromatographs. The detection limit of various aniline derivatives after preconcentration of 14 to 20 times is between 0.6 and 2.5 ng/L for SDS and 1.2 and 4.8 ng/L for HTAB.

Preconcentration increases the accuracy and precision of the analytical results.[25,26] Preconcentration of cadmium by MEUF is reported by Paulenova et al.[27] They use SDS as the surfactant and 8-hydroxyquinoline as the chelating agent. Cd first makes a complex in the extractant phase, and that phase is finally solubilized in the anionic micelle. It is observed that ultrafiltration recovery of Cd depends on the pH and ratio of concentration of metal and SDS. Also, recovery of Cd is a strong function of the concentration ratio of SDS and extractant. The optimum ratio of concentration of Cd to Cd + SDS is found to be 1:6 at pH 4.8. A pH around 5 is found to be the best value for extraction of Cd in this system.

Preconcentration and removal of iron (Fe(III)) from aqueous solutions are investigated by Pramauro et al.[28] by a mixed micellar system composed of nonionic surfactants $C_{12}H_8$ and the chelating agent salicylic acid. In the presence of nonionic surfactant, Brij 35, Fe(III) complexes with $C_{12}H_8$. The stoichiometry of the complexes is 1:1. Chelate formation is maximum at a pH range of 2.5 to 3.0. From the ng/L concentration range, a concentration of iron in the range of µg/L is achieved in the surfactant micelles in the retentate phase. pH 3.5 is found to be the most suitable for this system. More than 99% separation of iron is observed in the micellar phase. It is observed that higher hydrophobicity of the ligand leads to better retention of iron.

Hiraide and Itoh[29] demonstrated some limitation of preconcentration by MEUF and also provided a solution to that. They considered a case study of preconcentration of copper(II) ions. They used a feed with a concentration 50 µg/L of copper. A 20 g/L concentration of SDS was added to the solution and it was observed that at pH 2.0, after carrying out ultra-filtration by 10,000 cutoff membrane, more than 96% copper was trapped in the surfactant micelles. The surfactant-rich phase trapped with copper was injected with a graphite furnace atomic absorption spectrophotometer. A large interference in the determination of copper occurred due to the presence of surfactant micelles at 900 to 1,000°C. At higher temperature, the signals for copper and background intensity both decreased, and thereby the accuracy in the determination technique deteriorated. They also suggested an alternative to this problem. First, copper is chelated with ammonium pyrrolidinedithiocarbamate (APDC) that is hydrophobic. This hydrophobic complex is solubilized easily in SDS micelles. This surfactant-rich phase is then brought in contact with alumina particles in acidic environment (pH 2). Positively charged alumina adsorbs strongly the anionic surfactants. Alumina is then filtered by a 36-micron porous filter. Copper is desorbed from alumina by 4 M nitric acid solution, but the surfactants remain adsorbed on the alumina surface. The analysis is carried out easily without interference of the remaining surfactants. This method is applied to measure the concentration of copper in certified water samples collected from river and seawater. In the case of river water, the certified value is 1.35 ± 0.07 ng/ml, and with the developed method, it is 1.42 ± 0.05 ng/ml. In the case of seawater, the certified value is 0.297 ± 0.046 ng/ml, and with the developed method, it is 0.304 ± 0.011 ng/ml.

Pramauro et al.[30] have studied the preconcentrations of transition metal ions using MEUF. They have investigated the preconcentrations of five metal ions: nickel(II), copper(II), cobalt(II), manganese(II), and zinc(II). First, these metal ions are chelated with a complexing agent 1-(2-pyridylazole)-2-naphthol (PAN) and the hydrophobic complex is solubilized by nonionic surfactant Triton X-100. The micelles are separated by 10,000 molecular weight cutoff cellulose acetate membrane. The selected concentrations of various species are surfactant 0.02 M, PAN 6×10^{-6} M, and each metal ion 3 mg/L. At the normal pH of 7.0, retention of all five metal ions except calcium is more than 99%.

Guardia et al.[31] have carried out an investigation on the preconcentration of aluminum by MEUF. It is successfully made a complex with lumogallion and the complexed moiety is then solubilized in cationic cetylmethylammonium

TABLE 9.4

Summary of Preconcentration Studies Using MEUF

Analytes with Concentration	Surfactant with Concentration	Membrane System and Operating Conditions	Retention of Metal Ions	Reference
Aniline and its 17 derivatives in ppb concentration range	SDS (anionic) and HTAB (cationic) SDS: 0.1–0.3 M HTAB: 2 mM	10,000 MWCO, cellulose acetate membrane, $\Delta P =$ 300 kPa, batch cell pH: 2–3 for SDS; 7–8 for HTAB	SDS: 34–100% HTAB: 40–100%	24
Cadmium complexed with 8-hydroxyquinoline	SDS with cosurfactant n-butanol (10%) SDS: 0.006–0.03 M 8-HQ: 0.3–2.2 mM	20,000 MWCO, cellulose acetate membrane, $\Delta P =$ 400 kPa, batch cell pH: 24.8–5.3	>96%	27
Iron(III) (0.1mM) complexed with PAS-C_n	Hexadecyltrimethyl-ammonium nitrate (0.1 mM) PAS-C_n(1 mM)	10,000 MWCO, cellulose acetate membrane, $\Delta P =$ 300 kPa, batch cell pH: 3.5	>99.9	28
Copper(II) (50 $\mu g/l$)	SDS (20 g/L)	10,000 MWCO, cellulose acetate membrane, $\Delta P =$ not available, batch cell pH: 2.0	>96	29
Nickel(II), copper(II), cobalt(II), manganese(II), zinc(II) (3 mg/L, each)	Triton X-100 (0.02 M) with ligand PAN (6 ° 10^{-6} M)	10,000 MWCO, cellulose acetate membrane, ΔP =300 kPa, batch cell pH: 7.0	>99% for all except calcium	30
Aluminum in $\mu g/l$ concentration	Cetylmethyl-ammonium bromide (10 mM) Ligand:luminogallion (1 mM)	10,000 MWCO, cellulose acetate membrane, $\Delta P =$ 300 kPa, batch cell pH: 6.0	>98%	31

bromide (CTAB) micelles. The micelles are then removed by a 10,000 molecular weight cutoff ultrafiltration membrane. The suitable operating pH is found to be 5.0, and the concentrations of lumagallion and surfactant are 1 and 10 mM, respectively. Preconcentration of aluminum is achieved that is present initially at 6 mg/L. More than 98% retention of aluminum in the micelle phase is observed. The authors conclude that using this method, an Al concentration in the order of 30 μg/L can be analyzed. A summary of preconcentration studies using MEUF is presented in Table 9.4.

References

1. Pramauro, E., Prevot, H.A., Zelano, V., Gulmini, M., and Viscardi, G. 1996. Selective recovery of uranium(VI) from aqueous solutions using micellar ultrafiltration. *Analyst* 121: 1401–1405.
2. Ghezzi, L., Robinson, B.H., Secco, F., Tine, M.R., and Venturini, M. 2008. Removal and recovery of palladium (II) ions from water using micellar-enhanced ultrafiltration with a cationic surfactant. *Colloid. Surf. A Physicochem. Eng. Aspects* 329: 12–17.
3. Gwicanaa, S., Vorstera, N., and Jacobs, E. 2006. The use of a cationic surfactant for micellar-enhanced ultrafiltration of platinum group metal anions. *Desalination* 199: 504–506.
4. Akita, S., Yang, L., and Takecuchi, H. 1997. Micellar-enhanced ultrafiltration of gold(III) with nonionic surfactant. *J. Membr. Sci.* 133: 189–194.
5. Kirk, R.E., and Othmer, D.F. 1992. *Encyclopedia of chemical technology.* 4th ed. New York: John Wiley & Sons.
6. Bayraktar, E. 2001. Response surface optimization of the separation of DL-tryptophan using an emulsion liquid membrane. *Proc. Biochem* 37: 169–175.
7. Hossain, M.M. 2000. Mass transfer studies of amino acids and dipeptides in AOT-oleyl alcohol solution using a hollow fiber module. *Sep. Purif. Technol.* 18: 71–83.
8. Juang, R.S., and Wang, Y.Y. 2002. Amino acid separation of D2EHPA by solvent extraction of amino acids with cationic extractants. *J. Membr. Sci.* 207: 241–252.
9. Escudero, I., Ruiz, M.O., Benito, J.M., Cabezas, J.L., Dominguez, D., and Coca, J. 2006. Recovery of α-phenylglycine by micellar extractive ultrafiltration. *Chem. Eng. Res. Des.* 84(A7): 610–616.
10. Eyal, A.M., and Bressler, F. 1993. Industrial separation of carboxylic and amino acids by liquid membranes: Applicability, process considerations and potential advantages. *Biotechnol. Bioeng.* 41: 287–295.
11. Luque, S., Benito, J.M., and Coca, J. 2004. The importance of specification sheets of pressure driven membrane processes. *Filtrat. Sep.* 41: 24–28.
12. Sebasta, K., Farkas, J., and Horska, K. 1981. In *Microbial control of pests and plant diseases 1970–1980*, ed. H.D. Burges. New York: Academic. 249–281.
13. Ignoffo, C.M., and Gregory, B. 1972. Effect of *Bacillus thuringiensis. J. Econ. Entomol.* 63: 1987–1989.
14. de Barjac, H., Burgerjon, A., and Bonnefoi, A. 1966. The production of heat-stable toxin by nine serotypes of *Bacillus thuringiensis. J. Invertebr. Pathol.* 4: 537–538.
15. Benz, G. 1966. On the chemical nature of heat stable toxin of *Bacillus thuringiensis* Berliner in *Locusta Migration. J. Invertebr. Pathol.* 6: 381–383.
16. Kim, Y.T., and Hung, H.T. 1970. The β-exotoxin of *Bacillus thuringiensis* 1. Isolation and characterization. *J. Invertebr. Pathol.* 15: 100–108.
17. Tzeng, Y.M, Tsun, H.Y., and Chang, Y.N. 1999. Recovery of thuringiensin with cetylpyridinium chloride using micellar enhanced ultrafiltration process. *Biotechnol. Prog.* 15: 580–586.
18. Creagh, A.L., Hasenack, B.B.E., van der Padt, A., Sudholter, E.J.R., and vant Riet, K. 1994. Separation of amino acid enantiomers using micellar enhanced ultrafiltration. *Biotechnol. Bioeng.* 44: 690–698.

19. Overdevest, P.E.M., Bruin, T.J.M., Sudholter, E.J.R., et al. 2001. Separation of racemic mixture by ultrafiltration of enantioselective micelles. 1. Effect of pH on separation and regeneration. *Ind. Eng. Chem. Res.* 40: 5991–5997.
20. Overdevest, P.E.M., Schutyser, M.A.I., Bruin, T.J.M., et al. 2001. Separation of racemic mixture by ultrafiltration of enantioselective micelles. 2. (De)complexation kinetics. *Ind. Eng. Chem. Res.* 40: 5998–6003.
21. Overdevest, P.E.M., Hoenders, M.H.J., Riet, K. van't, van der Padt, A., and Kiurentjes, J.T.F. 2002. Enantiomer separation in a cascaded micellar enhanced ultrafiltration system. *AIChE J.* 48: 1917–1926.
22. Stalikas, C.D. 2002. Micelle-mediated extraction as a tool for separation and preconcentration in metal analysis. *Trends Anal. Chem.* 21: 343–355.
23. Pramauro, E., and Prevot, A.B. 1995. Solubilization in micellar systems. Analytical and environmental applications. *Pure Appl. Chem.* 67: 551–559.
24. Pramauro, E., Prevot, A.B., Savarino, P., Viscardi, G., Guardia, M. de la, and Cardelles, E.P. 1993. Preconcentration of aniline derivatives from aqueous solutions using micellar-enhanced ultrafiltration. *Analyst* 118: 23–27.
25. Mizuike, A. 1983. *Enrichment techniques for inorganic trace analysis.* Berlin: Springer.
26. Dunn, R.O., Scamehorn, J.F., and Christian, S.D. 1985. Use of micellar enhanced ultrafiltration to remove dissolved organics from aqueous streams. *Sep. Sci. Technol.* 20: 257–284.
27. Paulenova, A., Rajee, P., and Jezikova, M. 1998. Preconcentration of cadmium by MEUF in sodium dodecyl sulfate solutions. *J. Radioanal. Nucl. Chem.* 228: 119–122.
28. Pramauro, E., Bianco, A., Barni, E., Viscardi, G., and Hinze, W.L. 1992.Preconcentration and removal of iron(III) from aqueous media using micellar-enhanced ultrafiltration. *Colloids Surf.* 63: 291–300.
29. Hiraide, M., and Itoh, T. 2004. Ultrafiltration and alumina adsorption of micelles for the preconcentration of copper(II) in water. *Anal. Sci.* 20: 231–233.
30. Pramauro, E., Prevot, A.B., Zelano, V., Hinze, W.L., Viscardi, G., and Savarino, P. 1994. Preconcentration and selective metal ion separation using chelating micelles. *Talanta* 41: 1261–1267.
31. Guardia, M. de la, Peris-Cardelles, E., Morales-Rubio, A., Prevot, A.B., and Pramauro, E. 1993. Preconcentration of aluminium by micellar-enhanced ultrafiltration. *Anal. Chim. Acta* 276: 173–179.

Appendix: CMC Values of Some Surfactants

Compound Name	Solvent	Temp. (°C)	CMC (mM)	Reference
Anionic Surfactants				
$C_{10}H_{21}OCH_2COO^-Na^+$	0.1 M NaCl, pH 10.5	30	2.8	1
$C_{12}H_{25}COO^-K^+$	H_2O, pH 10.5	30	12	1
$C_9H_{19}CONHCH_2COO^-Na^+$	H_2O	40	38	2
$C_{11}H_{23}CONHCH_2COO^-Na^+$	H_2O	40	10	2
$C_{11}H_{23}CONHCH_2COO^-Na^+$	0.1 M NaOH (aq.)	45	3.7	3
$C_{11}H_{23}CON(CH_3)CH_2COO^-Na^+$	H_2O, pH 10.5	30	10	1
$C_{11}H_{23}CON(CH_3)CH_2COO^-Na^+$	0.1 M NaCl, pH 10.5	30	3.5	1
$C_{11}H_{23}CON(CH_3)CH_2CH_2COO^-Na^+$	H_2O, pH 10.5	30	7.6	1
$C_{11}H_{23}CON(CH_3)CH_2CH_2COO^-Na^+$	0.1 M NaCl, pH 10.5	30	2.7	1
$C_{11}H_{23}CONHCH(CH_3)COO^-Na^+$	0.1 M NaOH (aq.)	45	3.3	3
$C_{11}H_{23}CONHCH(C_2H_5)COO^-Na^+$	0.1 M NaOH (aq.)	45	2.1	3
$C_{11}H_{23}CONHCHCH(CH_3)_2COO^-Na^+$	0.1 M NaOH (aq.)	45	1.4	3
$C_{11}H_{23}CONHCHCH_2CH(CH_3)_2COO^-Na^+$	0.1 M NaOH (aq.)	45	0.58	3
$C_{13}H_{27}CONHCH_2COO^-Na^+$	H_2O	40	4.2	2
$C_{15}H_{31}CONHCHCH(CH_3)_2COO^-Na^+$	H_2O	25	1.9	4
$C_{15}H_{31}CONHCHCH_2CH(CH_3)_2COO^-Na^+$	H_2O	25	1.5	4
$C_8H_{17}SO_3^-Na^+$	H_2O	40	160	5
$C_{10}H_{21}SO_3^-Na^+$	H_2O	10	48	6
$C_{10}H_{21}SO_3^-Na^+$	H_2O	25	43	6
$C_{10}H_{21}SO_3^-Na^+$	H_2O	40	40	6
$C_{10}H_{21}SO_3^-Na^+$	0.1 M NaCl	10	26	6
$C_{10}H_{21}SO_3^-Na^+$	0.1 M NaCl	25	21	6
$C_{10}H_{21}SO_3^-Na^+$	0.1 M NaCl	40	18	6
$C_{10}H_{21}SO_3^-Na^+$	0.5 M NaCl	10	7.9	6
$C_{10}H_{21}SO_3^-Na^+$	0.5 M NaCl	25	7.3	6
$C_{10}H_{21}SO_3^-Na^+$	0.5 M NaCl	40	6.5	6
$C_{12}H_{25}SO_3^-Na^+$	H_2O	25	12.4	6
$C_{12}H_{25}SO_3^-Na^+$	H_2O	40	11.4	6
$C_{12}H_{25}SO_3^-Na^+$	0.1 M NaCl	25	2.5	6
$C_{12}H_{25}SO_3^-Na^+$	0.1 M NaCl	40	2.4	6
$C_{12}H_{25}SO_3^-Na^+$	0.5 M NaCl	40	7.9	6

Compound Name	Solvent	Temp. (°C)	CMC (mM)	Reference
Anionic Surfactants				
$C_{12}H_{25}SO_3^-Li^+$	H_2O	25	11.0	7
$C_{12}H_{25}SO_3^-NH_4^+$	H_2O	25	89.0	7
$C_{12}H_{25}SO_3^-K^+$	H_2O	25	9.3	7
$C_{14}H_{29}SO_3^-Na^+$	H_2O	40	2.5	5
$C_{16}H_{33}SO_3^-Na^+$	H_2O	50	0.70	5
$C_8H_{17}SO_4^-Na^+$	H_2O	40	140	8
$C_{10}H_{21}SO_4^-Na^+$	H_2O	40	33.0	8
$C_{11}H_{23}SO_4^-Na^+$	H_2O	21	16.0	9
Branched $C_{12}H_{25}SO_4^-Na^+$	H_2O	25	14.2	10
Branched $C_{12}H_{25}SO_4^-Na^+$	0.1 M NaCl	25	3.8	10
$C_{12}H_{25}SO_4^-Na^+$	H_2O	25	8.2	11
$C_{12}H_{25}SO_4^-Na^+$	H_2O	40	8.6	12
$C_{12}H_{25}SO_4^-Na^+$	"Hard river" water ionic strength = 6.6 mM	25	>1.58	13
$C_{12}H_{25}SO_4^-Na^+$	0.1 M NaCl	21	5.6	9
$C_{12}H_{25}SO_4^-Na^+$	0.3 M NaCl	21	3.2	9
$C_{12}H_{25}SO_4^-Na^+$	0.1 M NaCl	25	1.62	9
$C_{12}H_{25}SO_4^-Na^+$	0.2 M NaCl (aq.)	25	0.83	14
$C_{12}H_{25}SO_4^-Na^+$	0.4 M NaCl (aq.)	25	0.52	14
$C_{12}H_{25}SO_4^-Na^+$	0.3 M urea	25	9.0	15
$C_{12}H_{25}SO_4^-Na^+$	H_2O-cyclohexane	25	7.4	16
$C_{12}H_{25}SO_4^-Na^+$	H_2O-octane	25	8.1	16
$C_{12}H_{25}SO_4^-Na^+$	H_2O-decane	25	8.5	16
$C_{12}H_{25}SO_4^-Na^+$	H_2O-heptadecane	25	8.5	16
$C_{12}H_{25}SO_4^-Na^+$	H_2O-cyclohexane	25	7.9	16
$C_{12}H_{25}SO_4^-Na^+$	H_2O-carbon tetrachloride	25	6.8	16
$C_{12}H_{25}SO_4^-Na^+$	H_2O-benzene	25	6.0	16
$C_{12}H_{25}SO_4^-Na^+$	0.1 M NaCl (ag.)-heptane	20	1.4	17
$C_{12}H_{25}SO_4^-Na^+$	0.1 M NaCl (aq.)-ethyl benzene	20	1.1	17
$C_{12}H_{25}SO_4^-Na^+$	0.1 M NaCl (aq.)-ethyl acetate	20	1.8	17
$C_{12}H_{25}SO_4^-Li^+$	H_2O	25	8.9	18
$C_{12}H_{25}SO_4^-K^+$	H_2O	40	7.8	19
$(C_{12}H_{25}SO_4^-)_2Ca^{2+}$	H_2O	70	3.4	20
$C_{12}H_{25}SO_4^-N(CH_3)_4^+$	H_2O	25	5.5	18
$C_{12}H_{25}SO_4^-N(C_2H_5)_4^+$	H_2O	30	4.5	21
$C_{12}H_{25}SO_4^-N(C_3H_7)_4^+$	H_2O	25	2.2	22

Compound Name	Solvent	Temp. (°C)	CMC (mM)	Reference
Anionic Surfactants				
$C_{12}H_{25}SO_4^-N(C_4H_9)_4^+$	H_2O	30	1.3	21
$C_{13}H_{27}SO_4^-Na^+$	H_2O	40	4.3	23
$C_{14}H_{29}SO_4^-Na^+$	H_2O	25	2.1	24
$C_{14}H_{29}SO_4^-Na^+$	H_2O	40	2.2	12
$C_{15}H_{31}SO_4^-Na^+$	H_2O	40	1.2	23
$C_{16}H_{33}SO_4^-Na^+$	H_2O	40	0.58	8
$C_{13}H_{27}CH(CH_3)CH_2SO_4^-Na^+$	H_2O	40	0.80	23
$C_{12}H_{25}CH(C_2H_5)CH_2SO_4^-Na^+$	H_2O	40	0.90	23
$C_{11}H_{23}CH(C_3H_7)CH_2SO_4^-Na^+$	H_2O	40	1.1	23
$C_{10}H_{21}CH(C_4H_9)CH_2SO_4^-Na^+$	H_2O	40	1.5	23
$C_9H_{19}CH(C_5H_{11})CH_2SO_4^-Na^+$	H_2O	40	2.0	23
$C_8H_{17}CH(C_6H_{13})CH_2SO_4^-Na^+$	H_2O	40	2.3	23
$C_7H_{15}CH(C_7H_{15})CH_2SO_4^-Na^+$	H_2O	40	3.0	23
$C_{12}H_{25}CH(SO_4^-Na^+)C_3H_7$	H_2O	40	1.7	8
$C_{10}H_{21}CH(SO_4^-Na^+)C_5H_{11}$	H_2O	40	2.4	8
$C_8H_{17}CH(SO_4^-Na^+)C_7H_{15}$	H_2O	40	4.3	8
$C_{18}H_{37}SO_4^-Na^+$	H_2O	50	2.3	24
$C_{10}H_{21}OC_2SO_3^-Na^+$	H_2O	25	15.9	6
$C_{10}H_{21}OC_2H_4SO_3^-Na^+$	0.1 M NaCl	25	5.5	6
$C_{10}H_{21}OC_2H_4SO_3^-Na^+$	0.5 M NaCl	25	2.0	6
$C_{12}H_{25}OC_2H_4SO_4^-Na^+$	H_2O	25	3.9	6
$C_{12}H_{25}OC_2H_4SO_4^-Na^+$	Hard river water ionic strength = 6.6 mM	25	0.81	6
$C_{12}H_{25}OC_2H_4SO_4^-Na^+$	0.1 M NaCl	25	0.43	25
$C_{12}H_{25}(OC_2H_4)_2SO_4^-Na^+$	0.5 M NaCl	25	0.13	6
$C_{12}H_{25}(OC_2H_4)_2SO_4^-Na^+$	H_2O	10	3.1	6
$C_{12}H_{25}(OC_2H_4)_2SO_4^-Na^+$	H_2O	25	2.9	6
$C_{12}H_{25}(OC_2H_4)_2SO_4^-Na^+$	H_2O	40	2.8	6
$C_{12}H_{25}(OC_2H_4)_2SO_4^-Na^+$	Hard river water ionic strength = 6.6 mM	25	0.55	25
$C_{12}H_{25}(OC_2H_4)_2SO_4^-Na^+$	0.1 M NaCl	10	0.32	6
$C_{12}H_{25}(OC_2H_4)_2SO_4^-Na^+$	0.1 M NaCl	25	0.29	6
$C_{12}H_{25}(OC_2H_4)_2SO_4^-Na^+$	0.1 M NaCl	40	0.28	6
$C_{12}H_{25}(OC_2H_4)_2SO_4^-Na^+$	0.5 M NaCl	10	0.11	6
$C_{12}H_{25}(OC_2H_4)_2SO_4^-Na^+$	0.5 M NaCl	25	0.10	6
$C_{12}H_{25}(OC_2H_4)_3SO_4^-Na^+$	0.5 M NaCl	40	0.10	6
$C_{12}H_{25}(OC_2H_4)_4SO_4^-Na^+$	H_2O	50	2.0	24
$C_{12}H_{25}(OC_2H_4)_4SO_4^-Na^+$	H_2O	50	1.3	25
$C_{16}H_{33}(OC_2H_4)_5SO_4^-Na^+$	H_2O	25	0.025	26

Compound Name	Solvent	Temp. (°C)	CMC (mM)	Reference
Anionic Surfactants				
$C_8H_{17}CH(C_6H_{13})CH_2(OC_2H_4)_5SO_4^-Na^+$	H_2O	25	0.086	26
$C_6H_{13}OOCCH_2SO_3^-Na^+$	H_2O	25	170	27
$C_8H_{17}OOCCH_2SO_3^-Na^+$	H_2O	25	66.0	27
$C_{10}H_{21}OOCCH_2SO_3^-Na^+$	H_2O	25	22.0	27
$C_8H_{17}OOC(CH_2)_2SO_3^-Na^+$	H_2O	30	46.0	28
$C_{10}H_{21}OOC(CH_2)_2SO_3^-Na^+$	H_2O	30	11.0	28
$C_{12}H_{25}OOC(CH_2)_2SO_3^-Na^+$	H_2O	30	2.2	28
$C_{14}H_{29}OOC(CH_2)_2SO_3^-Na^+$	H_2O	40	0.09	28
$C_4H_9OOCCH_2CH(SO_3^-Na^+)COOC_4H_9$	H_2O	25	200.0	29
$C_5H_{11}OOCH_2(SO_3^-Na^+)COOC_5H_{11}$	H_2O	25	53.0	29
$C_6H_{13}OOCH_2CH(SO_3^-Na^+)COOC_6H_{13}$	H_2O	25	14.0	27
$C_4H_9CH(C_2H_5)CH_2OOCCH_2CH(SO_3^-Na^+)COOCH_2CH(C_2H_5)C_4H_9$	H_2O	25	2.5	29
$C_8H_{17}OOCCH_2CH(SO_3^-Na^+)COOC_8H_{17}$	H_2O	25	0.91	30
$C_{12}H_{25}CH(SO_3^-Na^+)COOCH_3$	H_2O	13	2.8	31
$C_{12}H_{25}CH(SO_3^-Na^+)COOC_2H_5$	H_2O	25	2.25	31
$C_{12}H_{25}CH(SO_3^-Na^+)COOC_4H_9$	H_2O	25	1.35	31
$C_{14}H_{29}CH(SO_3^-Na^+)COOCH_3$	H_2O	23	0.73	31
$C_{16}H_{33}CH(SO_3^-Na^+)COOCH_3$	H_2O	33	0.18	31
$C_{11}H_{23}CON(CH_3)CH_2CH_2SO_4^-Na^+$	H_2O pH 10.5	30	8.9	1
$C_{11}H_{23}CON(CH_3)CH_2CH_2SO_4^-Na^+$	0.1 M NaCl, pH 10.5	30	1.6	1
$C_{12}H_{25}NHCOCH_2SO_4^-Na^+$	H_2O	35	5.2	32
$C_{12}H_{25}NHCO(CH_2)_3SO_4^-Na^+$	H_2O	35	4.4	32
p-$C_8H_{17}C_6H_4SO_3^-Na^+$	H_2O	35	15.0	33
p-$C_{10}H_{21}C_6H_4SO_3^-Na^+$	H_2O	50	3.1	33
$C_{10}H_{21}$-2-$C_6H_4SO_3^-Na^+$	H_2O	30	4.6	34
$C_{10}H_{21}$-3-$C_6H_4SO_3^-Na^+$	H_2O	30	6.1	34
$C_{10}H_{21}$-5-$C_6H_4SO_3^-Na^+$	H_2O	30	8.2	34
$C_{11}H_{23}$-2-$C_6H_4SO_3^-Na^+$	H_2O	35	2.5	35
$C_{11}H_{23}$-2-$C_6H_4SO_3^-Na^+$	Hard river water ionic strength = 6.6 mM	30	0.25	35
p-$C_{12}H_{25}C_6H_4 SO_3^-Na^+$	H_2O	60	1.2	33
$C_{12}H_{25}C_6H_4 SO_3^-Na^+$	0.1 M NaCl	25	0.16	36
$C_{12}H_{25}$-2-$C_6H_4 SO_3^-Na^+$	H_2O	30	1.2	35
$C_{12}H_{25}$-2-$C_6H_4 SO_3^-Na^+$	Hard river water ionic strength = 6.6 mM	30	0.063	35
$C_{12}H_{25}$-3-$C_6H_4 SO_3^-Na^+$	H_2O	30	2.4	34
$C_{12}H_{25}$-5-$C_6H_4 SO_3^-Na^+$	H_2O	30	3.2	35

Compound Name	Solvent	Temp. (°C)	CMC (mM)	Reference
Anionic Surfactants				
$C_{12}H_{25}$-2-C_6H_4 $SO_3^-Na^+$	Hard river water ionic strength = 6.6 mM	30	0.46	35
$C_{13}H_{27}$-2-$C_6H_4SO_3^-Na^+$	H_2O	35	0.72	35
$C_{13}H_{27}$-2-$C_6H_4SO_3^-Na^+$	Hard river water ionic strength = 6.6 mM	30	0.011	35
$C_{13}H_{27}$-5-$C_6H_4SO_3^-Na^+$	H_2O	30	0.76	35
$C_{13}H_{27}$-5-$C_6H_4SO_3^-Na^+$	Hard river water ionic strength = 6.6 mM	30	0.083	35
$C_{16}H_{33}$-7-$C_6H_4SO_3^-Na^+$	H_2O	45	0.051	37
$C_{16}H_{33}$-7-$C_6H_4SO_3^-Na^+$	0.051 M HCl	45	0.0032	37

Compound	Solvent	Temp. (°C)	CMC (mM)	Reference
Cationic Surfactants				
$C_8H_{17}N^+(CH_3)_3Br^-$	H_2O	25	140.0	5
$C_{10}H_{21}N^+(CH_3)_3Br^-$	H_2O	25	68.0	5
$C_{10}H_{21}N^+(CH_3)_3Br^-$	0.1 M NaCl	25	42.7	38
$C_{10}H_{21}N^+(CH_3)_3Cl^-$	H_2O	25	68.0	39
$C_{12}H_{25}N^+(CH_3)_3Br^-$	H_2O	25	16.0	5
$C_{12}H_{25}N^+(CH_3)_3Br^-$	Hard river water ionic strength = 6.6 mM	25	12.6	25
$C_{12}H_{25}N^+(CH_3)_3Br^-$	0.01 M NaBr	25	12.0	40
$C_{12}H_{25}N^+(CH_3)_3Br^-$	0.1 M NaBr	25	4.2	40
$C_{12}H_{25}N^+(CH_3)_3Br^-$	0.5 M NaBr	31.5	1.9	41
$C_{12}H_{25}N^+(CH_3)_3Cl^-$	H_2O	25	20.0	42
$C_{12}H_{25}N^+(CH_3)_3Cl^-$	0.1 M NaCl	25	5.76	38
$C_{12}H_{25}N^+(CH_3)_3Cl^-$	0.5 M NaCl	31.5	3.8	41
$C_{12}H_{25}N^+(CH_3)_3F^-$	0.5 M NaF	31.5	8.4	41
$C_{12}H_{25}N^+(CH_3)_3NO_3^-$	0.5 M NaNO$_3$	31.5	0.08	41
$C_{14}H_{29}N^+(CH_3)_3Br^-$	H_2O	25	3.6	43
$C_{14}H_{29}N^+(CH_3)_3Br^-$	Hard river water ionic strength = 6.6 mM	25	2.45	25
$C_{14}H_{29}N^+(CH_3)_3Br^-$	H_2O	40	4.2	44
$C_{14}H_{29}N^+(CH_3)_3Br^-$	H_2O	60	5.5	44
$C_{14}H_{29}N^+(CH_3)_3Cl^-$	H_2O	25	4.5	45

Compound	Solvent	Temp. (°C)	CMC (mM)	Reference
Cationic Surfactants				
$C_{16}H_{33}N^+(CH_3)_3Br^-$	H_2O	25	1.0	46
$C_{16}H_{33}N^+(CH_3)_3Br^-$	0.001 M KCl	30	0.5	47
$C_{16}H_{33}N^+(CH_3)_3Cl^-$	H_2O	30	1.3	48
$C_{18}H_{37}N^+(CH_3)_3Br^-$	H_2O	40	0.34	49
$C_{10}H_{21}Pyr^+Br^-$	H_2O	25	44.0	50
$C_{10}H_{21}Pyr^+Br^-$	H_2O	25	63.0	51
$C_{10}H_{21}Pyr^+Br^-$	H_2O	25	21.0	50
$C_{12}H_{25}Pyr^+Br^-$	H_2O	10	11.7	52
$C_{12}H_{25}Pyr^+Br^-$	H_2O	25	11.4	52
$C_{12}H_{25}Pyr^+Br^-$	H_2O	40	11.2	52
$C_{12}H_{25}Pyr^+Br^-$	0.1 M NaBr	10	2.75	52
$C_{12}H_{25}Pyr^+Br^-$	0.1 M NaBr	25	2.75	52
$C_{12}H_{25}Pyr^+Br^-$	0.1 M NaBr	40	2.85	52
$C_{12}H_{25}Pyr^+Br^-$	0.5 M NaBr	10	1.07	52
$C_{12}H_{25}Pyr^+Br^-$	0.5 M NaBr	25	1.08	52
$C_{12}H_{25}Pyr^+Br^-$	0.5 M NaBr	40	1.16	52
$C_{12}H_{25}Pyr^+Cl^-$	H_2O	10	17.5	52
$C_{12}H_{25}Pyr^+Cl^-$	H_2O	25	17.0	52
$C_{12}H_{25}Pyr^+Cl^-$	H_2O	40	17.0	52
$C_{12}H_{25}Pyr^+Cl^-$	0.1 M NaCl	10	5.5	52
$C_{12}H_{25}Pyr^+Cl^-$	0.1 M NaCl	25	4.8	52
$C_{12}H_{25}Pyr^+Cl^-$	0.1 M NaCl	40	4.5	52
$C_{12}H_{25}Pyr^+Cl^-$	0.5 M NaCl	10	1.9	52
$C_{12}H_{25}Pyr^+Cl^-$	0.5 M NaCl	25	1.7	52
$C_{12}H_{25}Pyr^+Cl^-$	0.5 M NaCl	40	1.78	52
$C_{12}H_{25}Pyr^+I^-$	H_2O	25	5.3	52
$C_{13}H_{27}Pyr^+Br^-$	H_2O	25	5.3	50
$C_{14}H_{29}Pyr^+Br^-$	H_2O	25	2.7	50
$C_{14}H_{29}Pyr^+Cl^-$	H_2O	25	3.5	51
$C_{14}H_{29}Pyr^+Cl^-$	0.1 M NaCl	25	0.4	51
$C_{15}H_{31}Pyr^+Br^-$	H_2O	25	1.3	50
$C_{16}H_{33}Pyr^+Br^-$	H_2O	25	0.64	50
$C_{16}H_{33}Pyr^+Cl^-$	H_2O	25	0.90	53
$C_{18}H_{37}Pyr^+Cl^-$	H_2O	25	0.24	54
$C_{12}H_{25}N^+(C_2H_5)(CH_3)_2Br^-$	H_2O	25	14.0	55
$C_{12}H_{25}N^+(C_4H_9)(CH_3)_2Br^-$	H_2O	25	7.5	55
$C_{12}H_{25}N^+(C_6H_{13})(CH_3)_2Br^-$	H_2O	25	3.1	55
$C_{12}H_{25}N^+(C_8H_{17})(CH_3)_2Br^-$	H_2O	25	1.1	55
$C_{14}H_{29}N^+(C_2H_5)_3Br^-$	H_2O	25	3.1	43
$C_{14}H_{29}N^+(C_3H_7)_3Br^-$	H_2O	25	2.1	56, 43
$C_{14}H_{29}N^+(C_4H_9)_3Br^-$	H_2O	25	1.2	43

Compound	Solvent	Temp. (°C)	CMC (mM)	Reference
Cationic Surfactants				
$C_{10}H_{21}N^+(CH_2C_6H_5)(CH_3)_2Cl^-$	H_2O	25	39.0	57
$C_{12}H_{25}N^+(CH_2C_6H_5)(CH_3)_2Cl^-$	H_2O	25	8.8	58
$C_{14}H_{29}N^+(CH_2C_6H_5)(CH_3)_2Cl^-$	H_2O	25	2.0	58
$C_{12}H_{25}NH_2^+CH_2CH_2OH^-Cl^-$	H_2O	25	45.0	59
$C_{12}H_{25}NH_2^+(CH_2 CH_2OH)_2Cl^-$	H_2O	25	36.0	59
$C_{12}H_{25}NH_2^+(CH_2CH_2OH)_3Cl^-$	H_2O	25	25.0	59
$(C_{10}H_{21})_2N^+(CH_3)_2Br^-$	H_2O	25	1.85	55
$(C_{12}H_{25})_2N^+(CH_3)_2Br^-$	H_2O	25	0.18	55

Compound	Solvent	Temp. (°C)	CMC (mM)	Reference
Nonionic Surfactants				
$C_8H_{17}CHOHCH_2OH$	H_2O	25	2.3	60
$C_8H_{17}CHOHCH_2CH_2OH$	H_2O	25	2.3	60
$C_{10}H_{21}CHOHCH_2OH$	H_2O	25	0.18	60
$C_{12}H_{25}CHOHCH_2CH_2OH$	H_2O	25	0.013	60
n-Octyl-β-D-glucoside	H_2O	25	25.0	61
n-Decyl-α-D-glucoside	H_2O	25	22.0	62
n-Decyl-β-D-glucoside	H_2O	25	2.2	61
n-Dodecyl-α-D-glucoside	H_2O	60	0.072	63
Dodecyl-β-D-glucoside	H_2O	25	0.19	61
Decyl-β-D-maltoside	H_2O	25	2.0	62
Dodecyl-α-D-maltoside	H_2O	20	0.15	63
Dodecyl-β-D-maltoside	H_2O	25	0.15	62
$C_{12.5}H_{26}$ alkylpolyglucoside (degree of polym., 1.3)	H_2O	20	0.022	63
Tetradecyl-α-D-maltoside	H_2O	20	0.015	63
Tetradecyl-α-D-maltoside	H_2O	20	0.015	64
n-$C_4H_9(OC_2H_4)_6OH$	H_2O	20	800	64
n-$C_4H_9(OC_2H_4)_6OH$	H_2O	40	710	64
$(CH_3)_2CHCH_2(OC_2H_2)_6OH$	H_2O	20	910	64
$(CH_3)_2CHCH_2(OC_2H_2)_6OH$	H_2O	40	850	64
n-$C_6H_{13}(OC_2H_4)_6OH$	H_2O	20	74.0	64
n-$C_6H_{13}(OC_2H_4)_6OH$	H_2O	40	52.0	64
$(C_2H_5)_2CHCH_2(OC_2H_4)_6OH$	H_2O	20	100	64
$(C_2H_5)_2CHCH_2(OC_2H_4)_6OH$	H_2O	40	87.0	64
$C_8H_{17}OC_2H_4OH$	H_2O	25	4.9	65
$C_8H_{17}(OC2H4)_3OH$	H_2O	25	7.5	66
$C_8H_{17}(OC2H4)_5OH$	H_2O	25	9.2	67
$C_8H_{17}(OC2H4)_5OH$	0.1 M NaCl	25	5.8	67

Compound	Solvent	Temp. (°C)	CMC (mM)	Reference
Nonionic Surfactants				
$C_8H_{17}(OC_2H_4)_6OH$	H_2O	25	9.9	66
$(C_3H_7)_2CHCH_2(OC_2H_4)_6OH$	H_2O	20	23.0	64
$C_{10}H_{21}(OC_2H_4)_4OH$	H_2O	25	0.68	68
$C_{10}H_{21}(OC_2H_4)_5OH$	H_2O	25	7.6	69
$C_{10}H_{21}(OC_2H_4)_6OH$	H_2O	25	0.90	66
$C_{10}H_{21}(OC_2H_4)_6OH$	Hard river water ionic strength = 6.6 mM	25	0.87	25
$C_{10}H_{21}(OC_2H_4)_8OH$	H_2O	15	1.4	70
$C_{10}H_{21}(OC_2H_4)_8OH$	H_2O	25	1.0	70
$C_{10}H_{21}(OC_2H_4)_8OH$	H_2O	40	0.76	70
$(C_4H_9)_2CHCH_2(OC_2H_4)_6OH$	H_2O	20	3.1	64
$(C_4H_9)_2CHCH_2(OC_2H_4)_9OH$	H_2O	20	3.2	64
$C_{11}H_{23}(OC_2H_4)_8OH$	H_2O	15	0.40	70
$C_{11}H_{23}(OC_2H_4)_8OH$	H_2O	25	0.30	70
$C_{11}H_{23}(OC_2H_4)_8OH$	H_2O	40	0.23	71
$C_{12}H_{25}(OC_2H_4)_2OH$	H_2O	10	0.038	71
$C_{12}H_{25}(OC_2H_4)_2OH$	H_2O	25	0.033	71
$C_{12}H_{25}(OC_2H_4)_2OH$	H_2O	40	0.032	71
$C_{12}H_{25}(OC_2H_4)_3OH$	H_2O	10	0.063	71
$C_{12}H_{25}(OC_2H_4)_3OH$	H_2O	25	0.052	71
$C_{12}H_{25}(OC_2H_4)_3OH$	H_2O	40	0.056	71
$C_{12}H_{25}(OC_2H_4)_4OH$	H_2O	10	0.082	71
$C_{12}H_{25}(OC_2H_4)_4OH$	H_2O	25	0.064	71
$C_{12}H_{25}(OC_2H_4)_4OH$	Hard river water ionic strength = 6.6 mM	40	0.059	71
$C_{12}H_{25}(OC_2H_4)_4OH$		25	0.048	25
$C_{12}H_{25}(OC_2H_4)_5OH$	H_2O	10	0.090	71
$C_{12}H_{25}(OC_2H_4)_5OH$	H_2O	25	0.064	71
$C_{12}H_{25}(OC_2H_4)_5OH$	H_2O	40	0.059	71
$C_{12}H_{25}(OC_2H_4)_5OH$	0.1 M NaCl	25	0.064	67
$C_{12}H_{25}(OC_2H_4)_5OH$	0.1 M NaCl	40	0.059	67
$C_{12}H_{25}(OC_2H_4)_6OH$	H_2O	30	0.087	72
$C_{12}H_{25}(OC_2H_4)_6OH$	Hard river water ionic strength = 6.6 mM	25	0.069	25
$C_{12}H_{25}(OC_2H_4)_7OH$	H_2O	10	0.121	71
$C_{12}H_{25}(OC_2H_4)_7OH$	H_2O	25	0.082	71
$C_{12}H_{25}(OC_2H_4)_7OH$	H_2O	40	0.073	71
$C_{12}H_{25}(OC_2H_4)_7OH$	0.1 M NaCl (aq.)	25	0.08	73
$C_{12}H_{25}(OC_2H_4)_8OH$	H_2O	10	0.156	71

Compound	Solvent	Temp. (°C)	CMC (mM)	Reference
Nonionic Surfactants				
$C_{12}H_{25}(OC_2H_4)_8OH$	H_2O	25	0.11	71
$C_{12}H_{25}(OC_2H_4)_8OH$	H_2O	40	0.093	71
$C_{12}H_{25}(OC_2H_4)_8OH$	H_2O-cyclohexane	25	0.10	72
$C_{12}H_{25}(OC_2H_4)_8OH$	H_2O-heptane	25	0.1	72
$C_{12}H_{25}(OC_2H_4)_8OH$	H_2O-hexadecane	25	0.1	72
$C_{12}H_{25}(OC_2H_4)_9OH$	H_2O	23	0.10	74
$C_{12}H_{25}(OC_2H_4)_{12}H$	H_2O	23	0.14	75
6-branched-$C_{13}H_{27}(OC_2H_4)_5OH$	H_2O	25	0.28	67
6-branched-$C_{13}H_{27}(OC_2H_4)_5OH$	H_2O	40	0.21	67
$C_{13}H_{27}(OC_2H_4)_5OH$	H_2O	25	0.49	67
$C_{13}H_{27}(OC_2H_4)_5OH$	0.1 M NaCl (aq.)	25	0.021	67
$C_{13}H_{27}(OC_2H_4)_5OH$	H_2O	15	0.032	76
$C_{13}H_{27}(OC_2H_4)_8OH$	H_2O	25	0.027	76
$C_{13}H_{27}(OC_2H_4)_8OH$	H_2O	40	0.02	76
$C_{14}H_{29}(OC_2H_4)_6OH$	H_2O	25	0.01	66
$C_{14}H_{19}(OC_2H_4)_6OH$	Hard river water ionic strength = 6.6 mM	25	0.07	25
$C_{14}H_{19}(OC_2H_4)_8OH$	H_2O	15	0.011	70
$C_{14}H_{19}(OC_2H_4)_8OH$	H_2O	25	0.009	70
$C_{14}H_{19}(OC_2H_4)_8OH$	H_2O	40	0.0072	70
$C_{14}H_{19}(OC_2H_4)_8OH$	Hard river water ionic strength = 6.6 mM	25	0.01	25
$C_{15}H_{31}(OC_2H_4)_8OH$	H_2O	15	0.004	70
$C_{15}H_{31}(OC_2H_4)_8OH$	H_2O	25	0.0035	70
$C_{15}H_{31}(OC_2H_4)_8OH$	H_2O	40	0.003	70
$C_{16}H_{33}(OC_2H_4)_6OH$	H_2O	25	0.0016	25
$C_{16}H_{33}(OC_2H_4)_6OH$	Hard river water ionic strength = 6.6 mM	25	0.002	25
$C_{16}H_{33}(OC_2H_4)_7OH$	H_2O	25	0.0017	77
$C_{16}H_{33}(OC_2H_4)_9OH$	H_2O	25	0.0021	77
$C_{16}H_{33}(OC_2H_4)_{12}OH$	H_2O	25	0.0023	77
$C_{16}H_{33}(OC_2H_4)_{15}OH$	H_2O	25	0.0031	77
$C_{16}H_{33}(OC_2H_4)_{21}OH$	H_2O	25	0.004	77
p-t-$C_8H_{17}C_6H_4O(C_2H_4O)_2H$	H_2O	25	0.13	78
p-t-$C_8H_{17}C_6H_4O(C_2H_4O)_3H$	H_2O	25	0.097	78
p-t-$C_8H_{17}C_6H_4O(C_2H_4O)_4H$	H_2O	25	0.13	78
p-t-$C_8H_{17}C_6H_4O(C_2H_4O)_5H$	H_2O	25	0.15	78
p-t-$C_8H_{17}C_6H_4O(C_2H_4O)_6H$	H_2O	25	0.21	78
p-t-$C_8H_{17}C_6H_4O(C_2H_4O)_7H$	H_2O	25	0.25	78

Compound	Solvent	Temp. (°C)	CMC (mM)	Reference
Nonionic Surfactants				
$p\text{-}t\text{-}C_8H_{17}C_6H_4O(C_2H_4O)_8H$	H_2O	25	0.28	79
$p\text{-}t\text{-}C_8H_{17}C_6H_4O(C_2H_4O)_9H$	H_2O	25	0.03	79
$p\text{-}t\text{-}C_8H_{17}C_6H_4O(C_2H_4O)_{10}H$	H_2O	15	0.033	79
$p\text{-}C_8H_{19}C_6H_4 (OC_2H_4O)_8OH$	H_2O	—	0.13	80
$C_9H_{19}C_6H_4(OC_2H_4O)_{10}OH$	H_2O	25	0.075	81
$C_9H_{19}C_6H_4(OC_2H_4O)_{10}OH$	3 M urea	25	0.10	81
$C_9H_{19}C_6H_4(OC_2H_4O)_{10}OH$	6 M urea	25	0.24	81
$C_9H_{19}C_6H_4(OC_2H_4O)_{10}OH$	1.5 M dioxane	25	0.10	81
$C_9H_{19}C_6H_4(OC_2H_4O)_{10}OH$	3 M dioxane	25	0.18	81
$C_9H_{19}C_6H_4(OC_2H_4O)_{31}OH$	H_2O	25	0.18	81
$C_9H_{19}C_6H_4(OC_2H_4)_{31}OH$	3 M urea	25	0.35	81
$C_9H_{19}C_6H_4(OC_2H_4)_{31}OH$	3 M urea	25	0.74	81
$C_8H_{19}C_6H_4(OC_2H_4)_{31}OH^g$	3 M dioxane	40	0.57	81
$C_6H_{13}OCH_2CH(CH_3)_2(OC_2H_4)_{9.9}OH$	H_2O	20	47.0	82
$C_6H_{13}OCH_2CH(CH_3)_3(OC_2H_4)_{9.7}OH$	H_2O	20	32.0	82
$C_6H_{13}OCH_2CH(CH_3)_4(OC_2H_4)_{9.9}OH$	H_2O	20	19.0	82
$C_7H_{15}OCH_2CH(CH_3)_3(OC_2H_4)_{9.7}OH$	H_2O	20	11.0	82
Sucrose monolaurate	H_2O	25	0.34	83
Sucrose monoleate	H_2O	25	0.005	83
$C_{11}H_{23}CO_N(C_2H_4OH)_2$	H_2O	25	0.0027	84
$C_{15}H_{31}CO_N(C_2H_4OH)_2$	H_2O	35	0.012	85
$C_{11}H_{23}CO_N(C_2H_4O)_4H$	H_2O	23	0.005	86
$C_{10}H_{21}CON(CH_3(CHOH)_4CH_2OH$	0.1 M NaCl	25	1.58	87
$C_{11}H_{23}CON(CH_3(CH)CH_2CHOHCH_2OH$	0.1 M NaCl	25	0.023	87
$C_{11}H_{23}CON(CH_3)CH_2(CHOH)_3CH_2OH$	0.1 M NaCl	25	0.33	87
$C_{11}H_{23}CON(CH_3)CH_2(CHOH)_3CH_2OH$	0.1 M NaCl	25	0.35	87
$C_{12}H_{25}CON(CH_3)CH_2(CHOH)_3CH_2OH$	0.1 M NaCl	25	0.078	87
$C_{13}H_{27}CON(CH_3CH_2(CHOH)_3CH_2OH$	0.1 M NaCl	25	0.015	87
$C_{10}H_{21}N(CH_3)CO(CHOH)_4CH_2OH$	H_2O	20	1.29	88
$C_{12}H_{25}N(CH_3)CO(CHOH)_4CH_2OH$	H_2O	20	0.146	88
$C_{14}H_{29}N(CH_3)CO(CHOH)_4CH_2OH$	H_2O	20	0.024	88
$C_{16}H_{33}N(CH_3)CO(CHOH)_4CH_2OH$	H_2O	20	0.0078	88
$C_{18}H_{37}N(CH_3)CO(CHOH)_4CH_2OH$	H_2O	20	0.003	88

Source: Partly obtained from Rosen, M.J., *Surfactants and Interfacial Phenomena*, John Wiley & Sons, Hoboken, NJ, 2004 with due permission by John Wiley & Sons; and Mukerjee, P., and Mysels, K.J., *Critical Micellar Concentration of Aqueous Systems*, National Standard Reference Data Series, National Bureau of Standards (U.S.), 1970.

References

1. Tsubone, K., and Rosen, M.J. 2001. Structural effect on surface activities of anionic surfactants having N-acyl-N-methylamide and carboxylate groups. *J. Colloid Interf. Sci.* 244: 394–398.
2. Desai, A., and Bahadur, P. 1992. Surface activity and micellar behaviour of sodium salts of N-acyl glycine, glycylglycine and diglycylglycine. *Tenside Surf. Det.* 29: 425–428.
3. Miyagishi, S., Asakawa, T., and Nishida, M. 1989. Hydrophobicity and surface activities of sodium salts of N-dodecanoyl amino acids. *J. Colloid Interf. Sci.* 131: 68–73.
4. Ohta, A., Ozawa, N., Nakashima, S., Asakawa, T., and Miyagishi, S. 2003. Krafft temperature and enthalpy of solution of N-acyl amino acid surfactants and their racemic modifications: Effect of the amino acid residue. *Colloid Polym. Sci.* 281: 363–369.
5. Klevens, H.B. 1948. Critical micelle concentrations as determined by refraction. *J. Phys. Chem.* 52: 130–148.
6. Dahanayake, M., Cohen, A.W., and Rosen, M.J. 1986. Relationship of structure to properties of surfactants. 13. Surface and thermodynamic properties of some oxyethylenated sulfates and sulfonates. *J. Phys. Chem.* 90: 2413–2418.
7. Mohle, L., Opitz, S., and Ohlench, U. 1993. Behavior of alkane sulphonates at interfaces. *Tenside Surf. Det.* 30: 104–109.
8. Evans, H.C. 1956. Alkyl sulphates. Part I. Critical micelle concentrations of the sodium salts. *J. Chem. Soc.* 78: 579–586.
9. Huisman, H.F. 1964. Light scattering of solutions of ionic detergents III. *K. Ned. Akad. Wet. Proc. Ser. B* 67: 388–406.
10. Varadaraj, R., Bock, J., Zushma, S., and Brons, N. 1992. Influence of hydrocarbon chain branching on interfacial properties of sodium dodecyl sulfate. *Langmuir* 8: 14–17.
11. Elworthy, P.H., and Mysels, K.J. 1966. The surface tension of sodium dodecyl-sulfate solutions and the phase separation model of micelle formation. *J. Colloid Interf. Sci.* 21: 331–347.
12. Flockhart, B.D. 1961. The effect of temperature on the critical micelle concentration of some paraffin-chain salts. *J. Colloid Sci.* 16: 484–492.
13. Rosen, M.J., and Song, L.D. 1996. Dynamic surface tension of aqueous surfactant solutions 8. Effect of spacer on dynamic properties of gemini surfactant solutions. *J. Colloid Interf. Sci.* 179: 261–268.
14. Sowada, R. 1994. The effect of electrolytes on the critical micelle concentration of ionic surfactants: The Corrin-Harkins equation. *Tenside Surf. Det.* 31: 195–199.
15. Schick, M.J. 1964. Effect of electrolyte and urea on micelle formation. *J. Phys. Chem.* 68: 3585–3592.
16. Rehfeld, S.J. 1967. Adsorption of sodium dodecyl sulfate at various hydrocarbon-water interfaces. *J. Phys. Chem.* 71: 738–745.
17. Vijayendran, B.R., and Bursh, T.P. 1979. Effect of oil phase polarity on the saturation adsorption of sodium lauryl sulfate at oil/water interfaces. *J. Colloid Interf. Sci.* 68: 383–386.

18. Mysels, K.J., and Princen, L.H. 1959. Light scattering by some laurylsulfate solutions. *J. Phys. Chem.* 63: 1696–1700.
19. Meguro, K., and Kondo, T. 1956. Interaction between dyes and surfactants. I. Coagulation and dispersion of dyes by surfactants. *J. Chem. Soc. Jpn. Pure Chem. Sect.* 77: 1236.
20. Corkill, J.M., and Goodman, J.F. 1962. Formation of micelles with mixed counterions. *Trans. Faraday Soc.* 58: 206–214.
21. Kondo, T., and Meguro, K. 1959. The interaction between dyes and surfactants. *Bull. Chem. Soc. Jpn.* 32: 267–271.
22. Mukerjee, P. 1967. The nature of the association equilibria and hydrophobic bonding in aqueous solutions of association colloids. *Adv. Colloid Interf. Sci.* 1: 242–275.
23. Gotte, E., and Schwuger, M.J. 1969. Überlegungen und Experimente zum Mechanismus des Waschprozesses mit primären Alkyl sulfaten. *Tenside* 6: 131–135.
24. Gotte, E. 1960. 3rd int. congr. surf. activity. *Cologne* 1: 45.
25. Rosen, M.J., Zhu, Y.-P., and Morrall, S.W. 1996. Effect of hard river water on the surface properties of surfactants. *J. Chem. Eng. Data* 41: 1160–1167.
26. Lange, H., and Schwuger, M.J. 1968. Mizellbildung und Krafft-Punkte in der homologen Reihe der Natrium-n-alkyl-sulfate einschließlich der ungeradzahligen Glieder. *Colloid Polym.* 223: 145–149.
27. Jobe, D.J., and Reinsborough, V.C. 1984. Micellar properties of sodium alkyl sulfoacetates and sodium dialkyl sulfosuccinates in water. *Can. J. Chem.* 62: 280–284.
28. Hikota, T., Morohara, K., and Meguro, K. 1970. The properties of aqueous solutions of sodium 2-sulfoethyl alkanoates and sodium alkyl β-sulfopropionates. *Bull. Chem. Soc. Jpn.* 43: 3913–3916.
29. Williams, E.F., Woodberry, N.T., and Dixon, J.K. 1957. Purification and surface tension properties of alkyl sodium sulfosuccinates. *J. Colloid Sci.* 12: 452–459.
30. Nave, S., Eastoe, J., and Penfold, J. 2000. What is so special about aerosol-OT? 1. Aqueous systems. *Langmuir* 16: 8733–8740.
31. Ohbu, K., Fujiwara, M., and Abe, Y. 1998. Physicochemical properties of *a*-sulfonated fatty acid esters. *Progr. Colloid Polym. Sci.* 109: 85–92.
32. Mizushima, H., Matsuo, T., Satah, N., Hoffman, H., and Grachner, D. 1999. Synthesis and properties of N-alkyl amide sulfates. *Langmuir* 15: 6664–6670.
33. Gershman, J.W. 1957. Physico-chemical properties of solutions of para long chain alkylbenzenesulfonates. *J. Phys. Chem.* 61: 581–584.
34. Van Oss, N.M., Daane, G.J., and Haandrikman, G. 1991. The effect of chemical structure upon the thermodynamics of micellization of model alkylarenesulfonates. III. Determination of the critical micelle concentration and the enthalpy of demicellization by means of microcalorimetry and a comparison with the phase separation model. *J. Colloid Interf. Sci.* 141: 199–217.
35. Zhu, Y.-P., Rosen, M.J., Morrall, S.W., and Tolls, J. 1998. Surface properties of linear alkyl benzene sulfonates in hard river water. *J. Surf. Det.* 1: 187–193.
36. Murphy, D.S., Zhu, Z.H., Yuan, X.Y., and Rosen, M.J. 1990. Relationship of structure to properties of surfactants. 15 isomeric sulfated polyoxyethylenated Guerbet alcohols. *J. Am. Oil Chem. Soc.* 67: 197–204.
37. Lascaux, M.P., Dusart, O., Granet, R., and Piekarski, S. 1984. Water-alkanes interfacial tensions of branched sodium hexadecylbenzenesulfonates. Influence of salinity. *J. Chim. Phys.* 81: 345–348.

38. Li, F., Rosen, M.J., and Sulthana, S.B. 2001. Surface properties of cationic gemini surfactants and their interaction with alkylglucoside or -maltoside surfactants *Langmuir* 17: 1037–1042.
39. McGowan, J.C., and Sowada, R. 1993. Characteristic volumes and properties of surfactants. *J. Chem. Technol. Biotechnol.* 58: 357–361.
40. Tanaka, A., and Ikeda, S. 1991. Adsorption of dodecyltrimethylammonium bromide on aqueous surfaces of sodium bromide solutions. *Colloids Surf.* 56: 217–228.
41. Anacker, E.W., and Ghose, H.M. 1963. Counterions and micelle size. I. Light scattering by solutions of dodecyltrimethylammonium salts. *J. Phys. Chem.* 67: 1713–1716.
42. Osugi, J., Sato, M., and Ifuku, N. 1965. Micelle formation of cationic detergent solution at high pressures. *Rev. Phys. Chem. Jpn.* 35: 32–37.
43. Lianos, P., and Zana, R. 1982. Micelles of tetradecyltrialkylammonium bromides with fluorescent probes. *J. Colloid Interf. Sci.* 88: 594–598.
44. Gorski, N., and Kalus, J. 2001. Temperature dependence of the sizes of tetradecyltrimethylammonium bromide micelles in aqueous solutions. *Langmuir* 17: 4211–4215.
45. Hoyer, H.W., and Marmo, A. 1961. The electrophoretic mobilities and critical micelle concentrations of the decyl-, dodecyl- and tetradecyltrimethylammonium chloride micelles and their mixtures. *J. Phys. Chem.* 65: 1807–1810.
46. Okuda, H., Imac, T., and Ikeda, S. 1987. The adsorption of cetyltrimethylammonium bromide on aqueous surfaces of sodium bromide solutions. *Colloids Surf.* 27: 187–200.
47. Varjara, A.K., and Dixit, S.G. 1996. Adsorption of alkyltrimethylammonium bromide and alkylpyridinium chloride surfactant series on polytetrafluoroethylene powder. *J. Colloid Interf. Sci.* 177: 359–363.
48. Ralston, A.W., Eggenberger, D.N., and Harwood, H.J. 1947. The electrical conductivities of long-chain quaternary ammonium chlorides containing hydroxyalkyl groups. *J. Am. Chem. Soc.* 69: 2095–2097.
49. Swanson-Vethamuthu, M., Feitosa, E., and Brown, W. 1998. Salt-induced sphere-to-disk transition of octadecyltrimethylammonium bromide micelles. *Langmuir* 14: 1590–1596.
50. Skerjanc, S., Kogej, K., and Cerar, J. 1999. Equilibrium and transport properties of alkylpyridinium bromides. *Langmuir* 15: 5023–5028.
51. Mehrian, T., de Keizer, A., Kortwegr, A.J., and Lyklema, J. 1993. Thermodynamics of micellization of n-alkylpyridinium chlorides. *Colloid Surf. A* 71: 255–267.
52. Rosen, M.J., Dahanayake, M., and Cohen, A.W. 1982. Relationship of structure to properties in surfactants. 11. Surface and thermodynamic properties of N-dodecyl-pyridinium bromide and chloride. *Colloids Surf.* 5: 159–172.
53. Hartley, G.S. 1938. The solvent properties of aqueous solutions of paraffin-chain salts. Part I. The solubility of trans-azobenzene in solutions of cetylpyridinium salts. *J. Chem. Soc.* 370: 1968–1975.
54. Evers, E.C., and Kraus, C.A. 1948. Properties of electrolytic solutions. XXXIV. Conductance of some long chain electrolytes in methanol—water mixtures at 25°C. *J. Am. Chem. Soc.* 70: 3049–3054.
55. Lianos, P., Lang, J., and Zana, R. 1983. Fluorescence probe study of the effect of concentration on the state of aggregation of dodecylalkyldimethylammonium bromides and dialkyldimethylammonium chlorides in aqueous solution. *J. Colloid Interf. Sci.* 91: 276–279.

56. Venable, R.L., and Nauman, R.V. 1964. Micellar weights of and solubilization of benzene by a series of tetradecylammonium bromides. The effect of the size of the charged head. *J. Phys. Chem.* 68: 3498–3503.
57. de Castillo, J.L., Czapkiewicz, J., Gonzalez Perez, A., and Rodriguez, J.R. 2000. Micellization of decyldimethylbenzylammonium chloride at various temperatures studied by densitometry and conductivity. *Colloid Surf. A* 166: 161–169.
58. Rodriguez, J.R., and Czapkiewicz, J. 1995. Conductivity and dynamic light scattering studies on homologous alkylbenzyldimethylammonium chlorides in aqueous solutions. *Colloid Surf. A* 101: 107–111.
59. Omar, A.M.A., and Abdel-Khalek, N.A. 1997. Cationic surfactants as flotation collectors: Surface and thermodynamic parameters of some cationic surfactants with special reference to their use as flotation collectors. *Tenside Surf. Det.* 34: 178–182.
60. Kwan, C.-C., and Rosen, M.J. 1980. Relationship of structure to properties in surfactants. 9. Syntheses and properties of 1,2- and 1,3-alkanediols. *J. Phys. Chem.* 84: 547–551.
61. Shinoda, K., Yamaguchi, T., and Hori, R. 1961. The surface tension and the critical micelle concentration in aqueous solution of β-D-alkyl glucosides and their mixtures. *Bull. Chem. Soc. Jpn.* 34: 237–241.
62. Aveyard, R., Binks, B.P., Chen, J., et al. 1998. Surface and colloid chemistry of systems containing pure sugar surfactant. *Langmuir* 14: 4699–4709.
63. Bocker, Th., and Thiem, J. 1989. Synthèse et détermination structurale d'alkyl glycosides [Synthesis and structural elucidation of alkyl glycosides]. *Tenside Surf. Det.* 26: 318–324.
64. Elworthy, P.H., and Florence, A.T. 1964a. Critical micelle concentrations and thermodynamics of micellization of synthetic detergents containing branched hydrocarbon chains. *Colloid Polym. Sci.* 195: 23–27.
65. Shinoda, K., Yamanaka, T., and Kinoshita, K. 1959. Surface chemical properties in aqueous solutions of non-ionic surfactants octyl glycol ether, -octyl glyceryl ether and octyl glucoside. *J. Phys. Chem.* 63: 648–650.
66. Corkill, J.M., Goodman, J.F., and Harrold, S.P. 1964. Thermodynamics of micellization of non-ionic detergents. *Trans. Faraday Soc.* 60: 202–207.
67. Varadaraj, R., Bock, J., Geissler, P., Zushma, S., Brons, N., and Colletti, T. 1991. Influence of ethoxylate distribution on interfacial properties of linear and branched ethoxylate surfactants. *J. Colloid Interf. Sci.* 147: 396–402.
68. Hudson, R.A., and Pethica, B.A. 1964. In *Chem. phys. appl. surf. active substances,* ed. J.Th.G. Overbeek, 631. Proceedings of the 4th International Congress, Vol. 4. New York: Gordon & Breach.
69. Eastoe, J., Dalton, J.S., Rogueda, P.G.A., Crooks, E.R., Pitt, A.R., and Simister, E.A. 1997. Dynamic surface tensions of nonionic surfactant solutions. *J. Colloid Interf. Sci.* 188: 423–430.
70. Meguro, K., Takasawa, Y., Kawahashi, N., Tabata, Y., and Ueno, M. 1981. Micellar properties of a series of octaethyleneglycol-n-alkyl ethers with homogeneous ethylene oxide chain and their temperature dependence. *J. Colloid Interf. Sci.* 83: 50–56.
71. Rosen, M.J., Cohen, A.W., Dahanayake, M., and Hua, X.Y. 1982a. Relationship of structure to properties in surfactants. 10. Surface and thermodynamic properties of 2-dodecyloxypoly(ethenoxyethanol)s, C12H25(OC2H4)xOH, in aqueous solution. *J. Phys. Chem.* 86: 541–545.

72. Rosen, M.J., and Murphy, D.S. 1991. Effect of the nonaqueous phase on interfacial properties of surfactants. 2. Individual and mixed nonionic surfactants in hydrocarbon/water systems. *Langmuir* 7: 2630–2635.

73. Corkill, J.M., Goodman, J.F., and Ottewill, R.H. 1961. Micellization of homogeneous non-ionic detergents. *Trans. Faraday Soc.* 57: 1627–1636.

74. Sulthana, S.B., Bhat, S.G.T., and Rakshit, A.K. 1997. Interfacial and thermodynamic properties of mixed anionic-nonionic binary surfactant system. *J. Surf. Sci. Technol.* 13: 20–31.

75. Lange, H. 1965. Untersuchungen über adsorbierte und gespreitete monomolekulare Schichten von Dodecylpolyglykoläthern auf Wasser. *Colloid Polym. Sci.* 201: 131–136.

76. Cosgrove, T. 2010. *Colloid science: Principles, methods and applications.* New York: John Wiley & Sons.

77. Elworthy, P.H., and MacFarlane, C.B. 1962. Surface activity of a series of synthetic non-ionic detergents. *J. Pharm. Pharmacol. Suppl.* 14: 100T–102T.

78. Crook, E.H., Fordyce, D.B., and Trebbi, G.F. 1963. Molecular weight distribution of nonionic surfactants. I. Surface and interfacial tension of normal distribution and homogeneous p,t-octylphenoxyethoxyethanols (ope's). *J. Phys. Chem.* 67: 1987–1994.

79. Crook, E.H., Trebbi, G.F., and Fordyce, D.B. 1964. Thermodynamic properties of solutions of homogeneous p,t-octylphenoxyethoxyethanols OPE_{1-10}. *J. Phys. Chem.* 68: 3592–3599.

80. Voicu, A., Elian, M., Balcan, M., and Anghel, D.F. 1994. Synthesis and micellar properties of some homogeneously ethoxylated nonlyphenols. *Tenside Surf. Det.* 31: 120–123.

81. Schick, M.J., and Gilbert, A.H. 1965. Effect of urea, guanidinium chloride, and dioxane on the cmc of branched-chain nonionic detergents. *J. Colloid Sci.* 20: 464–472.

82. Kucharski, S., and Chlebicki, J. 1974. The effect of polyoxypropylene chain length on the critical micelle concentration of propylene oxide—Ethylene oxide block copolymers. *J. Colloid Interf. Sci.* 46: 518–521.

83. Herrington, T.M., and Sahi, S.S. 1986. Temperature dependence of the micellar aggregation number of aqueous solutions of sucrose monolaurate and sucrose monooleate. *Colloids Surf.* 17: 103–113.

84. Rosen, M.J., Friedman, D., and Gross, M. 1964. A surface tension study of the interaction of dimethyldodecylamine oxide with potassium dodecanesulfonate in dilute aqueous solution. *J. Phys. Chem.* 68: 3219–3225.

85. Hayes, M.E., Ei-Emary, M., Schechter, R.S., and Wade, W.H. 1980. Surface tension and selected micellar solution thermodynamic properties of some (n,n-bis-2-hydroxyethyl) amides. *J. Disp. Sci. Technol.* 1: 297–321.

86. Kjellin, U.R.M., Claesson, P.M., and Linse, P. 2002. Surface properties of tetra (ethylene oxide) dodecyl amide compared with poly (ethylene oxide) surfactants. 1. Effect of the headgroup on adsorption. *Langmuir* 18: 6745–6753.

87. Zhu, Y.-P., Rosen, M.J., Vinson, P.K., and Morrall, S.W. 1999. Surface properties of N-alkanoyl-N-methyl glucamines and related materials. *J. Surf. Det.* 2: 357–362.

88. Burczyk, B., Wilk, K.A., Sokolowski, A., and Syper, L. 2001. Synthesis and surface properties of N-alkyl-N-methylgluconamides and N-alkyl-N-methyllactobionamides. *J. Colloid Interf. Sci.* 240: 552–558.

Index

T - #0427 - 071024 - C8 - 234/156/10 - PB - 9780367381271 - Gloss Lamination